Lecture Notes in Physics

Edited by H. Araki, Kyoto, J. Ehlers, München, K. Hepp, Zürich
R. Kippenhahn, München, D. Ruelle, Bures-sur-Yvette
H.A. Weidenmüller, Heidelberg, J. Wess, Karlsruhe and J. Zittartz, Köln
Managing Editor: W. Beiglböck

352

Kurt Bernardo Wolf (Ed.)

Lie Methods in Optics II

Proceedings of the Second Workshop
Held at Cocoyoc, Mexico
July 19–22, 1988

Springer-Verlag
Berlin Heidelberg GmbH

Editor

Kurt Bernardo Wolf
Instituto de Investigaciones en Matemáticas
Aplicadas y en Sistemas/Cuernavaca
Universidad Nacional Autónoma de México
Apdo. Postal 20-726, 01000 México DF, Mexico

ISBN 978-3-662-13768-0 ISBN 978-3-540-46878-3 (eBook)
DOI 10.1007/978-3-540-46878-3

Originally published by Springer-Verlag Berlin Heidelberg New York in 1989
Softcover reprint of the hardcover 1st edition 1989

2153/3140-543210 – Printed on acid-free paper

Lie Methods in Optics

THE SECOND WORKSHOP

COCOYOC, MÉXICO, JULY 18–22, 1988

After the first Lie Methods in Optics workshop,[1] informal contacts with the participants and readers of the proceedings volume[2] suggested the time was ripe for a second such gathering. In the intervening three and a half years, the applications of Lie algebras and groups to optics have spread and deepened. So it was recognized by Prof. E.C.G. Sudarshan, who agreed to co-chair the workshop. Unforeseen institutional problems prevented him from attending the event personally, but his line of work was well represented by Prof. R. Simon, from the Institute of Mathematical Sciences, Madras, India.

Lie methods have similarities and differences with the more traditional tools employed hitherto in light and magnetic optics; as a branch of mathematics, moreover, they are worthy of study by themselves. Chapter 1, written by PETER W. HAWKES, surveys the pre- and post-Lie landscapes of perturbation expansions. Information theory derives very succinctly from the Heisenberg algebra by metaplectic harmonic analysis. Chapter 2, by WALTER SCHEMPP, offers mathematical foundations for coherent optical computing in the form of parallel two-dimensional data compression, holographic image processing and interferometry, and neural architecture for pattern recognition.

Symbolic computation with Lie structures, extensively applied by many students and researchers originating from the University of Maryland, has yielded developments that enhance their use not only in magnetic and light optics, but in the very principles of perturbation expansions to high orders. The canonical integration and concatenation of Lie transformations are treated in Chapter 3 by ETIENNE FOREST and MARTIN BERZ, and Chapter 4 by ALEX J. DRAGT and LIAM M. HEALY. The Superconducting Super-Collider project has provided great impetus to the refinement of the theory and efficiency in the computation of aberration expansions of high rank in symplectic manifolds.

It is not lost upon us that optical computing is within the present technological ten-year horizon. Pending further developments in nonlinear and physical optics for gate circuitry —*photonics*—, the parallel and sequential

[1] CIFMO-CIO Workshop on Lie Methods in Optics, held at León, México, January 7–10, 1985.

[2] *Lie Methods in Optics*, Proceedings, Ed. by J. Sánchez-Mondragón and K.B. Wolf. Lecture Notes in Physics, Vol. 250 (Springer, Berlin, Heidelberg, 1986).

processing capabilities of optical computers will require appropriate mathematics. The operations of convolution and correlation of two signals performed by purely optical means is the theme of Chapter 5, contributed by MARK KAUDERER out of his former association with MOSHE NAZARATHY and JOSEPH W. GOODMAN. The mathematics comes from paraxial optics but the symplectic algebra is applied to space and time.

A foundation of wide-angle optics based on the Euclidean group of motions is offered in Chapter 6, by K.B. WOLF. It accommodates geometrical and Helmholtz wave optics, defines a *wavization* process corresponding to —but distinct from— Heisenberg-Weyl quantization, and is compatible with signal Fourier analysis. (It is the privilege of the scientists in the less-developed countries to be able to roam almost free of technological necessities.) Its true aim is at present to contribute to the æsthetics, the coherence, and the extensions of Lie methods to untrodden fields with promise, particularly polarization tomography. Dr. VLADIMIR I. MAN'KO, one of the contributors to the first volume, who could unfortunately not be present at the second meeting, has again participated in a choice theme here.[3] During February 1989, working together at IIMAS–CUERNAVACA, we were able to understand the relation between the Euclidean and Heisenberg-Weyl Lie optics. This is presented in Chapter 7. It was included in this volume because it establishes the bridge between the traditional turf of coherent state theory, paraxial optics including aberration expansions, and global 4π optics.

One topic that was severly missed in the first volume is the strand of polarization optics developed in India by N. Mukunda, R. Simon and E.C.G. Sudarshan. This time, unfortunately, it was personal tragedy that prevented their work from appearing here, and is the main factor in the delay in publication of these proceedings.

As in the first volume, the chapters were prepared by the workshop participants to be of lasting value to a wide audience, including mathematicians, group-theoretical physicists, as well as applied opticists. If Lie methods are to be used successfully, they should be not only understood by the prospective user, but provide him or her with the pleasure of revealing the hidden symmetries of Nature.

México, Summer 1989 K. B. Wolf

---

[3] In this case it was the local organizer's fault for changing the contemplated Winter meeting into a Summer one, according to correspondence with other participants, after the invitation was made —with due anticipation— through official channels.

Centro Internacional de Física y Matemáticas Aplicadas

In 1988 CIFMA initiated its international activities with a main program in Photonics, of which mathematical optics is a component. The Lie Methods in Optics Second Workshop took place in Cocoyoc, on July 19–22, 1988, with the informal discussion-room athmosphere of the first meeting (León, January 7–10, 1985). This volume collects the advances in the field.

CIFMA was created as a Civil Association in Cuernavaca, State of Morelos. It is intended that CIFMA develop in Mexico the manifold activities pioneered by the International Centre for Theoretical Physics in Trieste, Italy, with special attention to the perceived scientific and technological needs and strong points of this country. Within the Latin American region, the Cuernavaca center joins the network started decades ago by the Centro Latino Americano de Física (CLAF, Brazil) and Centro Internacional de Física (CIF, Colombia).

The persistent economic problems afflicting Mexico have crippled scientific institutions; national funds for scientific research have been particularly meager. For this reason we are most grateful to Dr. Salvador Malo, **Dirección General de Investigación Científica y Superación Académica** (1988), Ministry of Public Education, who also sponsored the first workshop, for the firm support of our endeavours.

The participation of Prof. R. Simon was possible through grants from the **Third World Academy of Sciences** (Trieste), and the **Fondo de Fomento Educativo BCH** (México DF) whom it is a pleasure to thank also for the visit of Dr. V.I. Man'ko, one of the contributors to this volume, co-beneficiary of the **Consejo Nacional de Ciencia y Tecnología**. The indispensable logistic support of Dr. Ignacio Méndez, **Instituto de Investigaciones en Matemáticas Aplicadas y en Sistemas**, and Dr. José Sarukhán, **Coordinación de la Investigación Científica** (1988, now Rector) of the **Universidad Nacional Autónoma de México** is gratefully acknowledged.

ABOUT THIS VOLUME: We may derive satisfaction from the state of scientific typography in Mexico. Isolated efforts of the past (*viz.* Lecture Notes in Physics volumes 189 [1983] and 250 [1986]) have yielded —at long last— a small but growing core of professional technical editors. The present volume was prepared in TEX by **José Luis Olivares Vázquez** and **Arturo Sánchez y Gándara**, in the workshop of the **Sociedad Mexicana de Física**. The SMF publishes the *Revista Mexicana de Física* and the *Boletín de la SMF*, which we modestly point out were, in 1986, the first Latin American science journals to be composed with the efficiency and quality standards allowed by Donald Knuth's system. At the Third National Meeting of the *Grupo de Usuarios de TEX* (Cuernavaca, January 1989) we counted over ninety volumes, including the *Enciclopedia de México*, being prepared or published. Since 1988, the President's Address to the Nation —together with all of its gory administrative annexes— are typeset in TEX. The spearhead of these developments was **Aurión Tecnología**, whose Director, **Armando Jinich**, we thank for shoring up the funds for the present volume. For much good software, we thank **Max Díaz** and **Miguel Navarro Saad**.

ARTURO SÁNCHEZ Y GÁNDARA

Contents

Participants

Martin Berz	Superconducting Super Collider Central Design Group c/o Lawrence Berkeley Laboratory Berkeley, CA 94720, USA
Joaquín Delgado	Departamento de Matemáticas Universidad Autónoma Metropolitana-Iztapalapa Av. Michoacán y Purísima s/n Iztapalapa DF
Etienne Forest	Exploratory Studies Group Accelerator Fusion Lawrence Berkeley Laboratory Berkeley, CA 94720, USA
Peter W. Hawkes	Laboratoire d'Optique Electronique Centre National de la Recherche Scientifique, BP 4347 31055 Toulouse Cedex, France
Liam M. Healy	Naval Research Laboratory Code 8242 Washington, DC 20375–5000, USA
Mark Kauderer	Division of Engineering The University of Texas San Antonio, TX 78285, USA
Enrique López Moreno	Departamento de Física Facultad de Ciencias Universidad Nacional Autónoma de México Ciudad Universitaria, México DF
R. Simon	The Institute of Mathematical Sciences Madras 600 113, India
Walter Schempp	Lehrstuhl für Mathematik der Universität Siegen D-5900 Siegen German Federal Republic
Kurt Bernardo Wolf	Instituto de Investigaciones en Matemáticas Aplicadas y en Sistemas-Unidad Cuernavaca Universidad Nacional Autónoma de México, Apdo. Postal 139-B 62191 Cuernavaca

GRADUATE STUDENTS

Noé Alcalá Ochoa
Gloria García Quirino — Programa Nacional de Opto-Electrónica, Centro de Investigaciones en Optica, Apdo. Postal 728, 37000 León, Guanajuato

OTHER CONTRIBUTORS TO THIS VOLUME

Alex J. Dragt — Department of Physics, University of Maryland, College Park, MD 20742, USA

Joseph W. Goodman — Department of Electrical Engeenering, Stanford University, Stanford, CA 94305, USA

Moshe Nazarathy — Physical Sciences Laboratory, Hewlett-Packard Laboratories, Palo Alto, CA 94304, USA

Vladimir I. Man'ko — P.N. Lebedev Institute of Physics, USSR Academy of Sciences, Leninsky Prospekt 52, 117924 Moscow, USSR

1

Lie methods in optics: an assessment

Peter W. Hawkes

ABSTRACT Many important results have now been obtained in particle optics by means of the Lie algebraic methods. We compare these methods with older procedures based on the use of characteristic functions and attempt to bring out the advantages and shortcomings of the various approaches. In particular, we discuss the interrelations between aberration coefficients, concatenation, and the role of computer algebra.

1.1 The arrival of Lie methods on the optical scene; a qualitative survey

It is never wise in a branch of science to conclude that the vein has been fully worked and that the mine can be abandoned. Even in so clearly defined and circumscribed a field as particle optics, the pessimists who have wailed that nothing remained to discover have repeatedly been proved wrong, most recently by the addition of Lie weapons to the arsenal that have been developed over the years. Understandably, it has taken a little while to see these methods in perspective, to appreciate their advantages, to comprehend their relation to the older techniques and to incorporate them into current thinking on aberration theory and the design of complex systems, for particle accelerators or, at lower energies, for the lens and deflector sequences needed in electron microscopy and in nanolithography. Some of the claims made for the Lie methods when these were very new now seem excessive but some of the reactions to them likewise seem regrettably unfriendly. One of the purposes of the present essay is to stand back, now that the dust has settled, and assess the present situation. Future studies of geometrical optics will have to present Lie methods as fully as the older techniques; how should they be incorporated and what features of them should be emphasized?

In pre-Lie times, the calculation of an optical system took a variety of forms, according to the constraints and requirements of the structure being investigated. At some stage, however, the Lagrangian was obtained as a series expansion in the terms of the field or potential expansions and in the off-axis coordinates and their derivatives. The aberration coefficients were then obtained either by solving a trajectory equation by the method of variation of parameters (trajectory method, Green's function) or by differentiation of an appropriate eikonal function. The great advantage of the latter is that any inter-relations between the primary aberration coefficients are obvious and do not have to be recognised (or left unrecognised) as they do with the trajectory method. This weakness of the latter, which was in other respects very attractive being simple to understand and grasp, was recognised and circumvented. Thus DE BROGLIE [7], for example, used the physical rule that rays are normal to wave surfaces to reduce the 12 primary aberration coefficients of magnetic round lenses permitted by symmetry to the familiar eight; MEADS [38] (recapitulated in detail in HAWKES [24]), used the invariance of the Poincaré integral to establish 28 relations between the 40 geometric aberration coefficients of quadrupoles generated by the trajectory method. Both were of course effectively using the Hamiltonian approach to optics, and recently, MEINEL AND THIEM [39] have used essentialy the same technique to find the smallest number of independent primary aberration coefficients of prisms.

In the case of systems of round lenses and quadrupoles, the calculations were almost exclusively limited to the primary (third order) geometrical aberrations and the primary chromatic aberrations. The only need for higher order coefficients arose in connection with aberration correction; if, for example, the primary spherical aberration could be cancelled by means of a corrector, how serious would the next higher order spherical aberration be? In other fields, prism optics for example, aberration calculations were pursued to a higher order (see WOLLNIK [57] and HAWKES AND KASPER [33] for references). We must emphasize that all these calculations were performed by hand, without the benefit of computer algebra, which was not used in particle optics until the late 1970s (HAWKES [30]; SOMA [50]).

In accelerator optics, the results of these calculations were invariably presented in terms of the coordination between incident and emergent rays: point and direction of arrival in terms of point and direction of departure. The planes of reference were frequently not conjugate. It was quickly realized that matrices were well-suited to characterize this mapping between input space and output space and the paraxial properties were soon written routinely in matrix form while a matricial notation was adopted for the aberration coefficients: $(x|x^3)$, for example.

Another aspect of aberration theory, important for the design of multi-element structures, is the dependence of the coefficients on the object and image distance at each element and on the positions of any pupils in the system. The latter is mainly of interest in low-energy devices, of course. Optical designers have wrestled with these questions for many years, the work of SMITH [48], [49] being an early milestone, unfortunately better known for its obscurity than for its contents, despite the proselytising efforts of PEGIS [40] and more recently, of VELZEL AND DE MEIJERE [55]. A major clarification of the situation came with the work of the van Heel school (STEPHAN [52]; BROWER [8], [9]) alas little published, but see VAN HEEL [34], [35].) Here matrix methods were developed and employed very adroitly to establish practical addition formulae for aberration coefficients expressed in terms of position in two planes (object and entrance pupil) and not in terms of position and gradient in a single plane. In 1963, VERSTER [56] gave an expression for the aberrations of electrostatic lenses, expressed as powers of position and gradient in the object plane instead of position in two planes, in terms of the working magnification and hence of object distance. His results, buried in a thesis dealing with "gauze lenses", passed virtually unnoticed for some years, until it was at last realised (HAWKES [24]) that all asymptotic aberrations could be written as polynomials in reciprocal magnification, m ($m = 1/M$ where M is the magnification). Explicit formulae for the coefficients occurring in these polynomials, which depend only on the lens geometry and excitation and not on the working conditions, were obtained soon after for round lenses and quadrupoles. Since the aberration structure must have the same form for a single optical element or a combination, it was obviously possible to express the polynomial coefficients for a doublet explicitly in terms of those for each of its members and these expressions were likewise obtained [25], [26], [27], [28].

Despite the far reaching nature of many of the results, there is a lack of generality in the work that we have been describing for although formulae were available for calculating higher order aberration coefficients by the eikonal method (STURROCK [53]), the amount of work involved in applying them was daunting and there did not seem to be much in the way of general laws to guide the intrepid. Very general studies of quite high-order aberrations had been made (BUCHDAHL [10], [11]; MARX [37]; FOCKE [18], [19] especially) but these did not seem very helpful confronted with the massive expressions to be manipulated if the fifth-order aberrations even of round lenses had to be calculated (*e.g.* HAWKES [22]). In 1971, ROSE AND PETRI [45], *cf.* [41], [42], [43], [44], presented a form of the perturbation theory that yields aberration coefficients at once sufficiently general and sufficiently manageable to be used to calculate secondary and, if necessary, even higher order aberration coefficients. These formulae were not systematically exploited at the time but now that computer algebra is becoming widespread, they could be used to generate aberration integrals for virtually all the common optical elements.

It was in this heterogeneous and ultra-sophisticated world of aberration studies, going back more than a century in the case of light optics, that Lie optics arrived, where it was received somewhat like the ghost of Hamlet's father: first,

It harrows me with fear and wonder;

then,

> *We do it wrong, being so majestical*
> *To offer it the show of violence;*
> *For it is, as the air, invulnerable,*
> *And our vain blows malicious mockery;*

and finally,

> *And therefore as a stranger give it welcome.*

The bewilderment of the aberrations "establishment" before the unfamiliar notation was gradually dissipated by Dragt's persuasive efforts to show how the new theory could complement and enrich the old and any hostility finally evaporated when DRAGT AND FOREST [15] showed that Lie methods were capable of revealing the sextupole aberration corrector proposed by CREWE [13] and analysed at the Giessen meeting on Charged Particle Optics (ROSE [43]).

Before giving this qualitative account a more mathematical form, it is well to realise that there is more than one strand to Lie Optics. Practical optics has, naturally enough, been most affected by efforts of Dragt and his school to show how useful the techniques can be for experimental design work, in charged particle optics especially but also in light optics. In parallel, we have the more abstract work of K.B. Wolf and various colleagues, the main theme of which is the intricacies of the group theoretical beams and joists that maintain the handsome façade (reviewed in CASTAÑOS, LÓPEZ-MORENO AND WOLF [12]). Other practical aspects of the Lie approach, such as the questions of radar design and of the underlying algebra of wave optics, much studied by SCHEMPP [47], are represented in [46] and elsewhere in this book and hence not considered further here. For a very readable presentation of the mathematical background, see [51].

1.2 Pre-Lie and post-Lie

In the opening section, we have given some idea of the situation in geometrical optics when the Lie methods arrived. Here, we shall consider some specific points and show where the older and the newer methods converge. We shall also raise an unsolved question and examine the role of computer algebra in these theories. The present discussion is confined to aberrations expressed in terms of the position and gradient (or momentum) in an "object" or "starting" plane, for this is the case to which the Lie methods have been principally applied. In order to keep the reasoning simple, we shall likewise limit the discussion to geometrical aberrations, through the chromatic aberrations of dispersive systems can be examined equally easily.

In traditional aberration theory, then, we set out from a modified Lagrangian, of the form

$$M = \sqrt{\Phi(1+\epsilon\Phi)(1+X'^2+Y'^2)} - \eta(A_X X' + A_Y Y' + A_z), \tag{2.1}$$

in which $\Phi(X,Y,z)$ denotes the electrostatic potential distribution and A_X, A_Y, A_z are the components of the magnetic vector potential $\mathbf{A}(X,Y,z)$. The constants ϵ and η denote $\epsilon = \frac{1}{2}em_0c^2 \approx 1\ \mathrm{MV}^{-1}$ and $\eta = \sqrt{e/2m_0} \approx 3 \times 10^5\ (\mathrm{C/kg})^{1/2}$. Series expansions for Φ and the components of $\mathbf{A}$ are inserted in Eq. (2.1) and a perturbation calculus is developed that yields aberration coefficients as integrals over various terms of M. (An excellent succint account is to be found in Rose [44] and for a full presentation see Sturrock [54] or Hawkes and Kasper [33].) Instead of using perturbation theory, however, we may alternatively write down the Euler equations of the variational law governing M, and solve these by the usual methods, which again yields aberration integrals but obscures the fact that some of these are intrinsically the same. Then it is necessary to establish separately any relations, which can be done by invoking the invariance of the Poisson brackets for position and momentum. This is exactly equivalent, as Rose [44] shows explicitly, to the symplectic condition,

$$\mathbf{M}^\top \mathbf{J}_S \mathbf{M} = \mathbf{M}\mathbf{J}_S\mathbf{M}^\top = \mathbf{J}_S, \tag{2.2}$$

where $\mathbf{M}$ is the Jacobi matrix:

$$\frac{\partial(w, p^*, w^*, p)}{\partial(w_0, p_0^*, w_0^*, p_0)},$$

in which w, w_0 are position coordinates, p, p_0 momentum coordinates and the asterisk denotes a complex conjugate. Matrix transposition is indicated by $\mathbf{M}^\top$. $\mathbf{J}_S$ is the antisymmetric matrix:

$$\mathbf{J}_S = \begin{pmatrix} \mathbf{J} & 0 \\ 0 & \mathbf{J} \end{pmatrix}, \qquad \mathbf{J} = \begin{pmatrix} 0 & 1 \\ -1 & 0 \end{pmatrix}. \tag{2.3}$$

Thus, since all the interrelations between aberrations that are revealed automatically in perturbation calculations can be obtained by applying the invariance of the Poisson brackets to the results of a calculation by the trajectory method, it seems clear that application of the symplectic condition will yield exactly the same information.

The foregoing reasoning seems to exhaust the situation for primary aberrations but a knowledge of these is not always sufficient. If we need to calculate the secondary aberrations —the fifth-order aberrations of systems of round lenses and quadrupoles, for example— the argument becomes more complicated, for the integral expressions for these aberration coefficients consist of two parts: a contribution similar in appearance to the third-order coefficients coming from the truly fifth-order terms in the expansion of M;

and a mixed term arising from the interplay of the primary (third-order) aberrations and the quadratic terms in the M-expansion. Inter-relations between the contributions to each coefficient from the first part are again obvious but any such relations arising from the second part are extremely difficult to recognize. Indeed, a desultory and largely unpublished discussion concerning the very existence of such relations has been going on for many years (*e.g.* [20]). Admittedly, a thorough exploration of the integral expressions would certainly have revealed these putative relations but the labour involved seemed disproportionate to the result to be published. The situation changes when we return to the trajectory method (variation of parameters) and rely on the condition of symplecticity to reveal inter-relations. In the work of WOLLNIK AND BERZ [58], which is concerned with systems with a curved axis and hence having quadratic primary geometric aberrations as well as suitable measures of temporal and chromatic aberrations, sets of relations are established for each order. For the lowest order, the relations involve only terms of that order; for the next higher order, terms of this order and of the lowest occur; and in general, for each order, terms of all lower orders appear in the relations. This reminds us of the relations obtained by MEADS [38] already mentioned, which involve not only the third-order coefficients of quadrupoles but also their (paraxial) cardinal elements, focal lengths in particular.

We must pause to emphasize the importance of this result, and of its practical exploitation by WOLLNIK AND BERZ [58], for if the academically most interesting question that we ask of a new theory is, what new insights does it afford, close on the heels of this is a second less disinterested one: does it render any of the calculations easier or less laborious? So far as this question of inter-relations is concerned, therefore, the use of the symplectic condition is of no particular interest for establishing relations between primary aberrations, since they do not even need to be established when the perturbation eikonal is used, being obvious from the outset. For secondary or higher-order aberrations, however, the situation is reversed and we shall be well advised to establish any inter-relations by means of the symplectic condition as they may well not be immediately apparent in the perturbation expressions (for evidence of this last remark, see for example HAWKES [22], LI AND NI [36], or AI AND SZILAGYI [1]). An examination of the published formulae for the fifth-order aberrations of round electron lenses in the light of the predictions of the symplectic condition will be published separately. The relations revealed by applying the symplectic condition are again invaluable for situations so complex that is easier to obtain the numerical values of aberration coefficients not by evaluating aberration integrals but by very exact ray tracing, followed by repeated partial differentiation of the image space ray. Methods of calculating field distributions in three dimensions are now highly developed and the ray equations can be solved very accurately. Independent knowledge of any inter-relations to be expected clearly provides a useful check.

We now examine another aspect of optical design and again compare the results offered by the two theories, or rather, the two ways of exploiting Hamiltonian optics. Computer-aided design of optical columns, or more generally optical structures, is a laborious task. One of the reasons being that it is difficult, without much thankless brute-force computing, to identify the sources of large contributions to the aberrations of the overall system. This problem is ubiquitous: at low energies, it is only recently that practically realisable projector-lens combinations free of isotropic and anisotropic distortions have been found; at high energies, many problems in the design of accelerator structures that must satisfy a variety of constraints remain to be solved; and between these two extremes, the symmetry laws that must be obeyed in well-designed imaging energy analysers and time-of-flight spectrometers have only become clear in the past few years.

At low energies, an approach has been proposed that depends on a detailed analysis of aberration structure (HAWKES [28], [29], [32]). For asymptotic aberrations, that is, those that relate rays and ray manifolds in two spaces, which it is convenient to refer to as the object (space of incident rays) and image (space of emergent rays) spaces, we can write each coefficient as a polynomial in reciprocal magnification m. For the primary isotropic aberrations of round lenses we have

$$\begin{pmatrix} C \\ K \\ A \\ F \\ D \end{pmatrix} = \mathbf{Q} \begin{pmatrix} m^4 \\ m^3 \\ m^2 \\ m \\ 1 \end{pmatrix} \tag{2.4}$$

in which C denotes spherical aberration, K coma, A astigmatism, F field curvature and D distortion. The matrix $\mathbf{Q}$, eight of the elements of which are zero, contains only integrals over the field or potential in the lens and first order quantities, for focal lenghts.

These aberration coefficients are themselves elements in the matrix that connects object and image space. If we limit the discussion to third-order geometrical aberrations, we have

$$\mathbf{u}_m = \mathbf{M}\mathbf{u}_0, \tag{2.5}$$

in which $\mathbf{u}_m$ and $\mathbf{u}_0$ are column vectors corresponding to a pair of conjugate planes between which the magnification is $\mathbf{M}$. The elements of $\mathbf{u}$ are of the form

$$\mathbf{u}^{\mathsf{T}} = (u, u', ur^2, u'r^2, uV, u'V, u\theta^2, u'\theta^2), \tag{2.6}$$

in which $u = x + iy$, $u' = x' + iy'$, $r^2 = x^2 + y^2$, $V = xx' + yy'$, and $\theta^2 = x'^2 + y'^2$. The matrix $\mathbf{M}$ has the block form

$$\mathbf{M} = \begin{pmatrix} \mathbf{M}_1 & \mathbf{M}_2 \\ \mathbf{M}_3 & \mathbf{M}_4 \end{pmatrix}, \tag{2.7}$$

where $\mathbf{M}_1$ has the paraxial form

$$\mathbf{M}_1 = \begin{pmatrix} M & 0 \\ -1/f & m \end{pmatrix}, \tag{2.8}$$

and $m = 1/M$, f is the focal length and the refractive index is assumend to be the same on both sides of the lens. $\mathbf{M}_3$ is null and $\mathbf{M}_4$ encodes the rules for adding aberrations, as we shall see. The most important matrix for our present purposes is $\mathbf{M}_2$, the aberration matrix, which has two rows and (for isotropic aberrations) six columns. The upper row contains the aberrations of position and the lower row, those of gradient or momentum. The latter prove to be linear combinations of the former, plus one new term. Calculation of the aberrations of a doublet and hence by iteration of a multiplet with any number of members, is thus reduced to a matrix multiplication. It is not difficult, manually or with the aid of one of the computer algebra languages, to write down the aberration coefficients of the doublet in terms of those of the individual members and hence to identify any undesirable large contributions or unstable cancellations (differences between large and nearly equal terms, for example). Thus, formally we might write

$$\mathbf{u}_p = \mathbf{N}\mathbf{u}_m = \mathbf{N}\mathbf{M}\mathbf{u}_0 = \mathbf{P}\mathbf{u}_0, \tag{2.9}$$

where

$$\mathbf{P} = \begin{pmatrix} \mathbf{P}_1 & \mathbf{P}_2 \\ 0 & \mathbf{P}_4 \end{pmatrix} = \begin{pmatrix} \mathbf{N}_1 & \mathbf{N}_2 \\ 0 & \mathbf{N}_4 \end{pmatrix} \begin{pmatrix} \mathbf{M}_1 & \mathbf{M}_2 \\ 0 & \mathbf{M}_4 \end{pmatrix}, \tag{2.10}$$

giving

$$\mathbf{P}_2 = \mathbf{N}_1\mathbf{M}_2 + \mathbf{N}_2\mathbf{M}_4. \tag{2.11}$$

We can, however, go further than this. We know that the elements of the matrices $\mathbf{P}_2$, $\mathbf{M}_2$ and $\mathbf{N}_2$ can each be written as a polynomial in the reciprocal magnifications p, m and n respectively. In consecuence, it must be possible to write the coefficients of the polynomials (matrix $\mathbf{Q}$ in Eq. (2.4) for $\mathbf{P}_2$ in terms of those occurring in $\mathbf{M}_2$ and $\mathbf{N}_2$. This has been done but a somewhat surprising obstacle was encountered, which had the effect of emasculating the power of computer algebra. When we perform the matrix multiplication and attempt to extract the expressions for the polynomial coefficients, we find that the aberration coefficients in $\mathbf{P}_2$ appear not to possess the polynomial dependence on p that we know they must exhibit. Close scrutiny reveals that these coefficients consist of terms having the anticipated polynomial dependence and others that do not but which prove to form combinations that can be shown to vanish identically. That they do vanish is by no means always obvious and could easily be overlooked if we did not know what to look for. It is for this reason that a computer algebra language is, so to say, brought up short and, at the present time at least, it is necessary to complete such a calculation by hand, which in turn requires some degree of familiarity with the form of the coefficients; it requires too a well-developed faculty for mathematical pattern recognition.

Let us now consider how this problem is solved by the Lie method. That the latter is well adapted to handle it was noticed very early (DRAGT [13]) and regarded as one of the attractions of the Lie procedure, though for primary aberrations at least, it is not superior to the matrix multiplication described above (HAWKES [30]). As this is a very important point and there may well be some residual misunderstanding, we quote and comment two remarks from DRAGT AND FOREST's survey of 1986 [15]. Discussing the use of characteristic functions, they write:

> This same circumstance makes it difficult to find explicitly the characteristic function for a compound optical system even when the characteristic function for each of its component parts is already known.

A little latter, in their preliminary comments on the Lie algebraic approach to the problem of characterizing charged particle optical systems, they write:

> Moreover, it has the feature that the relation between initial and final quantities is always given in *explicit* form. In particular, it is possible to represent the action of each separate element of a compound optical system, including all departures from linear matrix optics, by a certain operator. These operators can then be concatenated by following well-defined rules to obtain a resultant operator that characterizes the entire system. That is, the use of Lie algebraic methods provides an operator extension of the linear matrix methods to the general case. Additionally, their use, as is the case with characteristic functions, facilitates the treatment of symmetries in an optical system and the classification of aberrations. Finally, it is believed that the Lie algebraic approach will greatly simplify the calculation of high-order aberrations.

Much of the effort has indeed been expended in light optics on solving the problem of calculating a characteristic function for a compound system given that of its components; from the numerous studies, it emerges that the angle characteristic offers a reasonable way of obtaining a solution and with its aid, FOCKE (1951 [18]; 1965 [19]) has derived explicit addition formulae for the aberration coefficients of a system of several partial systems (Eqs. 3.47 and 3.48 of [19]). Nevertheless, straightforward concatenation of operators would certainly be a welcome alternative and we feel that the matrix multiplication already described and the Lie procedure offer equivalent though formally very different ways of doing this. In the matrix multiplication technique, each optical element is characterized by a matrix, which can be decomposed into blocks, each with a physical interpretation: a paraxial block matrix, a primary aberration block matrix, block matrices coding the addition rules for aberrations, and null matrices.

Manual manipulation of matrices becomes prohibitively laborious beyond the primary aberrations but such manipulations are trivial with computer algebra. In the Lie picture, each optical element is represented by a transfer map, $\mathcal{M}$, which relates object and image space just like the matrices we have been discussing. For round lenses, for example, this map has the form of a product of exponentials

$$\mathcal{M} = \exp(: f_2 :)\exp(: f_4 :)\exp(: f_6 :)\cdots \tag{2.12}$$

(Eq. 26 of [16]) in which f_2 corresponds to paraxial behavior and f_4, $f_6 \ldots$ to the primary, secondary and higher-order aberrations, respectively. The colon notation indicates that the associated quantity is a Lie operator,

$$: f : g = [f, g] \tag{2.13}$$

and $[f, g]$ is the Poisson bracket of classical mechanics. The combined transfer map $\mathcal{M}_h$ for two elements, characterized by $\mathcal{M}_f$ and $\mathcal{M}_g$, is obtained from

$$\mathcal{M}_h = \mathcal{M}_f \mathcal{M}_g \tag{2.14}$$

and since it must be possible to write $\mathcal{M}_h$ in the form

$$\mathcal{M}_h = \exp(: h_2 :)\exp(: h_4 :)\exp(: h_6 :)\cdots, \tag{2.15}$$

our task is to obtain the h_j in terms of f_j and g_j, $j = 2, 4, 6 \ldots$. This can be accomplished with the aid of the Baker-Campbell-Hausdorff formula, which tells us that the product of exponentials of non-commuting operators (here $: f :$ and $: g :$) can indeed be written as the exponential of some new operator (here $: h :$) and that the latter is given by the sum of the original operators plus terms having the form of simple and multiple commutators of these operators. As in the case of the matrix formulation, the terms of lowest order can be obtained by hand calculation but beyond these, computer algebra is indispensable.

We have seen that, for the systematic design of complex systems, the fact that aberration coefficients can be separated into a part depending only on the geometry and excitation of each optical element and a part depending only on the relation of each optical element to its neighbours (aberration polynomials in reciprocal magnification) is important. We have seen too that the task of obtaining recurrence relations for the polynomial coefficients is not as straightforward as we could wish, for some of the power of computer algebra is lost.

Can the Lie methods help here? We have as yet no answer to this question but it merits further consideration, for the way in which the operators for different optical elements are combined is very different from simple matrix multiplication. Can the superfluous terms that have to be recognized and extracted by human vigilance in the matrix calculation be made to emerge naturally in the Lie concatenation step? It is far from obvious that they can, for the aberrations are characterized in a way that does not seem to lend itself to such a result. Nevertheless, the presence of these intrusive terms is unwelcome and renders the calculation of recurrence formulae more laborious than one would like and it is certainly worth enquiring whether any other technique is superior.

We conclude this section with a comment on the ease of implementation of the various approaches. Even for relatively simple situations, both are laborious: for example, the calculation of the relativistically correct form of the primary (third-order) geometrical aberrations of round electrostatic lenses is a formidable task, unless computer algebra is used, and is even more thankless if all the many partially integrated forms are required [29]. One of the main practical reasons why computer algebra languages were first developed was, however, the need to perform power series expansions involving large numbers of terms and they are therefore extremely well-adapted to aberration calculations; alternatively, of course, special-purpose languages can be developed (see BERZ [2], [3], BERZ AND WOLLNIK [4], BERZ *et al.* [5] and BERZ [this volume]). Computer algebra programs are generally not difficult to write nor are they usually very long, for the operations to be performed are rather few in number; the number of lines devoted to sorting the results and presenting them in a reasonably legible manner may well form a substantial part of the code.

The use of these computer algebra languages in particle optics has, however, been mainly confined to the study of relatively complicated situations by the Lie algebraic methods (MACSYMA and SMP for various versions for MARYLIE). The reason for this is partly historical, partly fortuitous. In electron optics, where simple field models are inadequate, the aberration integrals for the basic optical elements —round lenses and quadrupoles and pure deflectors— have long been known and those for combined lenses and deflectors, although obtained more recently, were also derived manually. The transfer matrices for prisms were likewise established in the 1960s and early 1970s [57]. As we have mentioned, there was litle incentive to go beyond the primary aberrations of lenses intended for such instruments as microscopes. It is in the field of accelerator optics and, to some extent, mass-spectrometer optics, that it was desirable to calculate the higher-order behaviour of charged-particle beams and the new Lie-optical programs, supported by computer algebra, are being developed for this task. If it were worth devoting a similar effort to writing the necessary algebra programs for calculating the higher-order aberrations of short lenses and quadrupoles, say, then the eikonal-based methods and Lie methods would again be seen to require a comparable effort to obtain equivalent, but complementary, results. Nevertheless, as soon as the expresions become so complicated that the inter-relations between individual members are no longer obvious, it will always be worth establishing these by invoking the invariance of the Poisson brackets or, equivalently, the symplectic condition.

1.3 Concluding remarks

The object of this paper has been to compare and contrast the three families of methods of studying optical systems: the trajectory method [33], [57], the use of characteristic functions or eikonals [21], [33], [44], [53], [54], and the Lie algebraic methods [13], [17]. We agree entirely with the comment of DRAGT *et al.* [17] that

> there is nothing that can be computed using Lie algebraic methods that could not, in principle, be computed using Hamiltonian perturbation theory or, indeed, Newton's equations. However, in practice, because of their concise and modular nature, the use of Lie algebraic methods has made it possible to obtain many results well beyond those currently available by any other method.

We must, however, stress that matrix methods of representing particle optical elements permitting concatenation and hence optimization already exist and, at least in the present state of the theory, are superior to the Lie methods in one respect. In situations where the optical elements can be represented by a field or potential function that remains constant within the element and vanishes outside it, it is reasonable to relate input and output quantities and join these with matrices representing the drift spaces between endfaces; this approximation can of course be improved by introducing the notion of fringing fields but these are effectively localized at the endfaces. The Lie approach in accelerator design has naturally concentrated on this situation. At lower energies, the approximation is not even useful for crude calculations as the lenses used typically have no region of constant field. The matrices used thus map the quantities characterizing an incoming ray asymptote into those characterizing the corresponding outgoing asymptote and it is convenient that the quantities descriptive of the optical element (field distribution, excitation) can be dissociated from the mode of operation ("object" position, magnification) and that recurrence relations for these quantities can be established. This gives the designer a high degree of control over the parameters of the overall system and direct information about, for example, the sources of large contributions to individual aberrations.

In the Lie method and the eikonal method, we have encountered a misapprehension concerning the ease of use of the latter. We again quote from the recent review of DRAGT *et al.* [17]:

> Despite their general theoretical importance and frequent applicability, there are certain awkward features associated with the use of characteristic functions. For example, their use gives an admixture of initial and final quantities. These relations must be made explicit, by solving for the final quantities in terms of the initial quantities, before they can be applied.

In fact, this step, in which the final (image-space) quantities are solved for in terms of the initial (object-space) quantities is not at all awkward for the task is effectively performed when one or other of the perturbation eikonals is introduced. This is made clear in, for example, STURROCK [53] [54]; BORN AND WOLF [6] who describe in detail the use of the perturbation function that Schwarzschild called the "Seidel eikonal" (§5.2); in the survey by ROSE [44], see his Eqs. (119); and in [33], §§22 and 24, and especially Eqs. (22.36) and (24.29). The formulae that are used to calculate the aberration coefficients of a particular order give the latter immediately by partial differentiation of functions involving the field distribution and lower order approximations to the trajectories with respect to an initial (object-space) coordinate.

This said, it is in no sense my intention to provoke a polemic between the proponents of Lie optical methods and some (perhaps already fictitious) old guard of eikonal lovers: having myself used characteristic functions for more than a quarter of century, I am delighted to find that familiarity with the Lie approach has enabled me to solve at least one long-standing problem. I am convinced that these different approaches must henceforward be regarded as complementary, some problems being easier to treat by one rather than the other and in practice, better software often being available for one than for the other. Moreover, for complex cases, it is certainly wise to use both methods in order to check the results. It is therefore highly desirable that optical designers should be familiar with both and that the two approaches should be described fully in future texts on aberrations.[1]

1.4 REFERENCES

[1] K.-c. Ai and M. Szilagyi, Fifth-order relativistic aberration theory for electromagnetic focusing systems incluiding spherical cathode lenses, *Optik* **79**, 33–40 (1988).

[2] M. Berz, The method of power series tracking for the mathematical description of beam dynamics, *Nucl. Instrum. Meth. Phys. Res.* **A 258**, 431–436 (1987).

[3] M. Berz, Differential algebraic description of beam dynamics to very high orders, *Lawrence Berkeley Laboratory Report* SSC-152 (1988).

[4] M. Berz and H.Wollnik, The program HAMILTON for the analytic solution of the equations of motion through fifth order, *Nucl. Instrum. Meth. Phys. Res.* **A 258**, 364–373 (1987).

[1]For a variety of reasons, it was not possible to include an account of the Lie algebraic methods in [33]; if one day a new edition should be called for, I shall follow my own advice and remedy matters.

[5] M Berz, H.C. Hoffman and H. Wollnik, COSY 5.0 —the fifth order code for corpuscular systems, *Nucl. Instrum. Meth. Phys. Res.* **A 258**, 402–406 (1987).

[6] M. Born and E. Wolf, *Principles of Optics* (Pergamon, Oxford, 1959; 6th ed., 1980).

[7] L. de Broglie, *Optique Electronique et Corpusculaire* (Hermann, Paris, 1950).

[8] W. Brouwer, *The use of matrix algebra in geometrical optics.* Dissertation, Delft (1957).

[9] W. Brouwer, *Matrix Methods in Optical Instrument Design* (Benjamin, New York and Amsterdam, 1964).

[10] H.A. Buchdahl, *Optical Aberration Coefficients* (Oxford University Press, Oxford, 1954 and Dover, New York, 1968).

[11] H.A. Buchdahl, *An Introduction to Hamiltonian Optics* (Cambridge University Press, Cambridge, 1970).

[12] O. Castaños, E. López-Moreno, and K.B. Wolf, Canonical transformations for paraxial wave optics, in *Lie Methods in Optics*, J. Sánchez-Mondragón and K.B. Wolf, Eds. (Springer-Verlag, Berlin, 1986), pp. 159–182.

[13] A.V. Crewe and D. Kopf, Limitations of sextupole correctors, *Optik* **56**, 391–399 (1980).

[14] A.J. Dragt, Lie algebraic theory of geometrical optics and optical aberrations, *J. Opt. Soc. Am.* **72**, 372–379 (1982).

[15] A.J. Dragt, Elementary and advanced Lie algebraic methods with applications to accelerator design, electron microscopes and light optics, *Nucl. Instrum. Meth. Phys. Res.* **A 258**, 339–354 (1987).

[16] A.J. Dragt and E. Forest, Lie algebraic theory of charged-particle optics and electron microscopes, *Advances in Electronics and Electron Physics* **67**, 65–120 (1986).

[17] A.J. Dragt, E. Forest and K.B. Wolf, Foundations of a Lie algebraic theory of geometrical optics, in *Lie Methods in Optics*, J. Sánchez-Mondragón and K.B. Wolf, Eds. (Springer-Verlag, Berlin, 1986), pp. 105–157.

[18] A.J. Dragt, F. Neri, G. Rangarajan, D.R. Douglas, L.M. Healy, and R.D. Ryne, Lie algebraic treatment of linear and nonlinear beam dynamics, *Ann. Rev. Nucl. Part. Sci.* **38**, 455–496 (1988).

[19] J. Focke, Die Realisierung der Herzbergerschen Theorie der Fehler fünfter Ordnung in rotationssymmetrischen Systemen mit einer Anwendung auf das Schmidtsche Spiegelteleskop, *Jenaer Jahrbuch* 89–119 (1951).

[20] J. Focke, Higher order aberration theory, *Progress in Optics* **4**, 1–36 (1965).

[21] A. Foster, *Correction of aperture aberrations in magnetic lens spectrometers.* Thesis, London (1968).

[22] W. Glaser, *Grundlagen der Electronenoptik* (Springer-Verlag, Vienna, 1952).

[23] P.W. Hawkes, The geometrical aberrations of general electron optical systems II. The primary (third order) aberrations of orthogonal systems, and the secondary (fifth order) aberrations of round systems, *Phil. Trans. Roy. Soc. London* **A 257**, 523–552 (1965).

[24] P.W. Hawkes, *Quadrupole Optics* (Springer-Verlag, Berlin, 1966).

[25] P.W. Hawkes, Asymptotic aberration coefficients, magnification and object position in electron optics, *Optik* **27**, 287–304 (1968).

[26] P.W. Hawkes, Asymptotic aberration integrals for round lenses, *Optik* **31**, 213–219 (1970).

[27] P.W. Hawkes, Asymptotic aberration integrals for quadrupole systems, *Optik* **31**, 302–314 (1970).

[28] P.W. Hawkes, The addition of round electron lens aberrations, *Optik* **31**, 592–599 (1970).

[29] P.W. Hawkes, The addition of quadrupole aberrations, *Optik* **32**, 50–60 (1970).

[30] P.W. Hawkes, Computer calculation of formulae for electron lens aberration coefficients, *Optik* **48**, 29–51 (1977).

[31] P.W. Hawkes, Lie algebraic theory of geometrical optics and optical aberrations: a comment, *J. Opt. Soc. Am.* **73**, 122 (1983).

[32] P.W. Hawkes, Aberration structure, *Nucl. Instrum. Meth. Phys. Res.* **A 258**, 462–465 (1987).

[33] P.W. Hawkes and E. Kasper, *Principles of Electron Optics*, 2 vols. (Academic Press, London and San Diego, 1989).

[34] A.C.S. van Heel, Calcul des aberrations du cinquième ordre et projects tenant compte de ces aberrations, in *La Théorie des Images Optiques*, P. Fleury, Ed. (Editions Revue d'Optique, Paris, 1949), pp. 32–67.

[35] A.C.S. van Heel, *Inleiding in de Optica* (Martinus Nijhoff, The Hague, 1964).

[36] Y. Li and W.-x. Ni, Relativistic fifth order geometrical aberrations of a focusing system, *Optik* **78**, 45–47 (1988).

[37] H. Marx, Theorie der geometrisch-optischen Bildfehler, in *Handbuch der Physik*, S. Flügge, Ed. (Springer, Berlin & New York, 1967), Vol. 29, Optische Instrumente, pp. 68–191.

[38] P. Meads, *The theory of aberrations of quadrupole focusing arrays*. Thesis, University of California (1963); UCRL-10807.

[39] R. Meinel and S. Thiem, A complete set of independent asymptotic aberration coefficients for deflection magnets, *Optik* **74**, 1–2 (1986).

[40] R. J. Regis, The modern development of Hamiltonian optics, *Progress in Optics* **1**, 1–29 (1961).

[41] H. Rose, Über die Berechnung der Bildfehler elektronenoptischer Systeme mit gerader Achse, *Optik* **27**, 466–474 and 497–514 (1968).

[42] H. Rose, Der Zusammenhang der Bildfehler-Koeffizienten mit den Entwicklungs-Koeffizienten des Eikonals, *Optik* **28**, 462–274 (1968/9).

[43] H. Rose, Correction of aperture aberrations in magnetic systems with threefold symmetry, *Nucl. Instr. Meth.* **187**, 187–199 (1981).

[44] H. Rose, Hamiltonian magnetic optics, *Nucl. Instrum. Meth. Phys. Res.* **A 258**, 374–406 (1987).

[45] H. Rose and U. Petri, Zur systematischen Berechnung elektronenoptischer Bildfehler, *Optik* **33**, 151–165 (1971).

[46] J. Sánchez-Mondragón and K.B. Wolf, Eds., *Lie Methods in Optics*, (Springer-Verlag, Berlin, 1986); Lecture Notes in Physics Vol. 250.

[47] W. Schempp, Analog radar signal design and digital signal processing —a Heisenberg nilpotent Lie group approach, in *Lie Methods in Optics*, J. Sánchez-Mondragón and K.B. Wolf, Eds. (Springer-Verlag, Berlin, 1986), pp. 1–27.

[48] T. Smith, The changes in aberrations when the object and stop are moved, *Trans. Opt. Soc. London* **23**, 311–322 (1921/2).

[49] T. Smith, The addition of aberrations, *Trans. Opt. Soc. London* **25**, 177–199 (1923/4).

[50] T. Soma, Relativistic aberration formulas for combined electric-magnetic focusing-deflection system, *Optik* **49**, 255–262 (1977).

[51] S. Steinberg, Lie series, Lie transformations, and their applications, in *Lie Methods in Optics*, J. Sánchez-Mondragón and K.B. Wolf, Eds. (Springer-Verlag, Berlin, 1986), pp. 45–103.

[52] W.G. Stephan, *Practische Toepassingen op het Gebied der algebraïsche Optica.* Dissertation, Delft (1947).

[53] P.A. Sturrock, Perturbation characteristic functions and their application to electron optics, *Proc. Roy. Soc. London* **A 210**, 269–289 (1951).

[54] P.A. Sturrock, *Static and Dynamic Electron Optics* (Cambridge University Press, Cambridge, 1955).

[55] C.H.F. Velzel and J.L.F. de Meijere, Characteristic functions and the aberrations of symmetric optical systems. I. Transverse aberrations when the eikonal is given. II. Addition of aberrations, *J. Opt. Soc. Am.* **A 5**, 246–250 and 251–256 (1988).

[56] J.L. Verster, On the use of gauzes in electron optics, *Philips Res. Repts.* **18**, 465–605 (1963).

[57] H. Wollnik, *Optics of Charged Particles* (Academic Press, Orlando & London, 1987).

[58] H. Wollnik and M. Berz, Relations between elements of transfer matrices due to the condition of symplecticity, *Nucl. Instrum. Meth. Phys. Res.* **A 238**, 127–140 (1985).

2

Holographic image processing, coherent optical computing, and neural computer architecture for pattern recognition

Walter Schempp

> ***At the end, I would like to express my belief that the holographic concept of Gabor is as fundamental as the general relativity theorem of Einstein, and it has to be explored further for a better understanding of nature in which we live.***
>
> Pál Greguss (1986)

ABSTRACT Metaplectic harmonic analysis is well matched with high resolution image processing. The metaplectic representation of the symplectic group and its twofold cover arises when the symplectic group is considered as a group of outer automorphisms of the irreducible linear representations of the Heisenberg two-step nilpotent Lie group. Starting with the Paley-Wiener theorem which forms the classical result for information-preserving sequential bandwidth compression, and its Stone-von Neumann-Segal analogue for the Heisenberg group which is at the basis of holographic reciprocity and coupling, the paper points out a unified metaplectic approach to signal geometry such as holographic image processing, coherent optical computing, and neural computer architecture for pattern recognition. Brief descriptions of hardware implementations are also included.

2.1 Introduction

> ***However, we find it interesting that the various descriptions, Doppler filtering, aperture synthesis, holography and cross-correlation, diverse as they are when described physically, become identical when formulated mathematically.***
>
> Emmett N. Leith (1978)

Holographic image processing is based on coherently scanning the combined amplitude and phase distributions out of interference patterns which are encoded in the holographic plane by mutually coherent wave mixing and linear mode superposition. Therefore, the hologram is a sort of associative memory which qualifies it as a main component of an optical computer.

Due to the very nature of the active interferometric method, the resolution of holographic imaging systems is independent of the range to the target and the velocity of the coherent radiation source. Therefore, the non-local holographic geometry can be described by the sub-Riemannian geometry of the three-dimensional real Heisenberg two-step nilpotent Lie group. The holographic transform H serves as a mathematical model of the basic operations of mixing and superposing of holographic image processing. The sesquilinear mapping H assigns to the elements of the complex tensor product vector space $L^2(\mathbf{R}) \otimes L^2(\mathbf{R})$ the matrix coefficient function of the linear Schrödinger representation of the Heisenberg group. It is the purpose of the present paper to study the quantum mechanical approach to holographic image processing including coherent optical computing, holographic reciprocity and coupling, linear and non-linear optical phase conjugation, electron holography, and transverse tomographic methods in high resolution imaging radar. In particular, synthetic-aperture imaging radar (SAR) like side-looking airborne radar (SLAR) and space shuttle imaging radar (SIR-A, SIR-B), are unified by the nilpotent harmonic analysis approach.

Specifically, the metaplectic representation gives rise to the implementation of a programmable real-time SAR processing architecture and also to the classical SAR processing architecture for coherently computing H. Finally, the pixel mapping associated with the toroidal compactification of the Heisenberg group gives rise to a classification of the planar holographic grids. The matching polynomials of the complete bipartite subgraphs form up to an attenuation factor the elementary holograms. Their coefficients count the synaptic interconnections between the binary formal neurons in a hierarchical manner. A hardware implementation of the hexagonal holographic grid by a silicon VLSI retina forms an opto-electronic neural computer which simulates the detection of oriented light-intensity edges of biological visual systems. This hybrid neural computer is useful in artificial pattern recognition, intelligent scanning, automatic analysis of radar images, robotics, and other artificial tasks of intelligent high resolution image processing.

2.2 Sequential data compression

I have never done anything "useful". No discovery of mine has made, or is likely to make, directly or indirectly, for good or ill, the least difference to the amenity of the world. I have helped to train other mathematicians, but mathematicians of the same kind as myself, and their work has been, so far at any rate as I have helped them to it, as useless as my own. Judged by all practical standards, the value of my mathematical life is nil; and outside mathematics it is trivial anyhow.

Godfrey H. Hardy (1877–1947)

Data compression can be defined as the reduction in the amount of signal space that must be allocated to a given message set. This signal space may be in a physical volume, for instance a data storage medium such as an optical disc or in a bounded interval of the electromagnetic spectrum, such as the bandwith required to transmit the given message.

There are two classes of data compression techniques: The information-preserving, or reversible, techniques and non-information-preserving, or irreversible, techniques. In the former, compression is achieved by reducing the entropy of the signal source. For instance, uniform sampling above the Nyquist rate of signals band-limited in frequency allows an ideal resampling procedure whereas degradation in the data by uniform amplitude quantization and fractal compression results in an irreversible loss of information.

The classical result concerning information-preserving sequential bandwidth compression is formed by the Paley-Wiener theorem (sometimes also called Plancherel-Pólya theorem). An analogue of the Paley-Wiener theorem for the Heisenberg two-step nilpotent Lie group, the Stone-von Neumann-Segal theorem, which at the basis of holographic reciprocity and coupling, will be established in Section 6 *infra*.

Let u denote an entire holomorphic function that satisfies an estimate of type

$$|u(w)| \leq Ce^{\pi|w|} \quad (w \in \mathbf{C}),$$

with a positive constant C. The classical Paley-Wiener theorem states that when u is square integrable along the real axis $\mathbf{R}$, then u can be reconstructed from a function $U \in L^2(T)$, where T denotes the one-dimensional compact torus group, by means of the finite Fourier cotransform of U. The complex-valued functions u that are square integrable along $\mathbf{R}$ and admit an entire holomorphic extension to $\mathbf{C}$ of exponential type $< \pi$ form the complex Paley-Wiener space $PW(\mathbf{C})$ which will be considered as a vector subspace of the standard complex Hilbert space $L^2(\mathbf{R})$.

The proof that the Fourier transform U of u vanishes outside the interval $(-1/2, +1/2)$ of $\mathbf{R}$ can be traced back to an idea of G.H. Hardy. It depends upon a complex contour integral argument, the Cauchy integral theorem, and the Phragmén-Lindelöf principle, which forms a far-reaching generalization of the maximum modulus principle of complex analysis. The complex contour is located in the upper (*resp.* lower) complex half-plane. The reflection $w \mapsto \bar{w}$ which transfers the path of the upper half-plane to the path in the lower half-plane has applications to non-linear optical phase conjugation (*cf.* Section 11 *infra*).

An inmediate consequence of the classical Paley-Wiener theorem is the fact that the complex Hilbert space $PW(\mathbb{C})$ admits a (unique) reproducing kernel function K given by

$$K(w, z) = \text{sinc}\,(w - \bar{z}),$$

where the function on the right hand side is defined according to the rule

$$\text{sinc}\; w = \begin{cases} \dfrac{\sin(\pi w)}{\pi w} & \text{for } w \neq 0, \\ 0 & \text{for } w = 0. \end{cases}$$

Notice that K is holomorphic in the first variable and anti-holomorphic in the second variable (see Schwartz [24]) and that every function $u \in PW(\mathbb{C})$ has the convolution representation

$$u(w) = \int_{\mathbb{R}} u(t)\,\text{sinc}\,(w - t)\,dt \qquad (w \in \mathbb{C}).$$

The classical Paley-Wiener theorem implies the uniform sampling theorem due to Whittaker-Nyquist-Shannon-Kotel'nikov which is a compressed or grid version of the preceding reproducing kernel property of the complex Hilbert space PW($\mathbb{C}$). Indeed, the uniform sampling theorem states for functions $u \in PW(\mathbb{C})$ the cardinal series representation by sinc-translates

$$u(w) = \sum_{\mu \in \mathbb{Z}} u(\mu)\,\text{sinc}\,(w - \mu) \qquad (w \in \mathbb{C}).$$

The resampling filter of the discrete convolution is defined by the prescription given above. Notice that the cardinal series for u is uniformly convergent in every horizontal strip of the complex plane $\mathbb{C}$. The fact that the sampling phase of band-limited signals is unspecified usually presents no serious problems when the sampling rate is well above the Nyquist rate. Oversampling, *i.e.*, sampling at a rate which is greater than the Nyquist rate, is typical in practice. For undersampled signals, however, the particular sampling phase may play a significant rôle. In any case, an unspecified phase implies a loss of information, which can be remedied by the holographic technique.

The uniform sampling theorem is at the basis of the Digital to Analogic (D/A)-conversion of sequential digital signal processing, for instance the Compact Disk Digital Audio (CD-A), which will be recognized in Section 10 *infra* as a one-dimensional hologram. Adopting this point of view, a D/A-converter can be considered as a decoding device.

For a survey of sampling theory, see the papers of BUTZER *et al.* [3], HIGGINS [12], and JERRI [14]. For an application of group theoretical methods, the reader is refered to the paper [20].

The resampling theory was first applied to computer graphics by CROW [4]. Notice that computer graphics, *i.e.*, image synthesis, and image processing, *i.e.*, image analysis, are in many ways inverse technologies. However, with growing graphics capabilities to create highly realistic synthetic scenes from compressed data, many image applications, such as remote sensor system performance (see Section 12, *infra*) and model-based computer vision (Section 13, *infra*), are now able to use sophisticated computer graphics technology. Fractal data compression based on chaotic dynamic systems theory (see BARNSLEY *et al.* [2]) and holding promise to realize $10^4 : 1$ compression ratios is part of the convergence of computer graphics and image processing. Presently the synergism is beggining to occur and to generate a new trend affecting the field of intelligent high resolution image processing.

2.3 Applications: CD-A, CD-ROM and CD-E

> ***Erasable memory systems for onboard data storage with capacities on the order of 1 terabit and data rates up 1.5 gigabit/second are expected to be required in the 1990's.***
>
> NASA (1988)

A digital audio signal is a discrete-time, discrete-amplitude representation of the original acoustic audio signal. Such a digital representation in binary coded format is more fault-tolerant with respect to recording, storage, and retrieval than an analogue representation. A discrete-time scale is realized by sampling a time-continuous signal. Due to the uniform sampling theorem, signals band-limited in frequency can be resampled without distortion. The accuracy of the resampled signal is determined by the accuracy with which the clock frequency can be established. A discrete amplitude is realized by measuring the amplitude of the samples and by rounding its value to a multiple of a chosen quantization unit. A consequence of the size of the quantization unit is that a restriction is imposed on the accuracy with which the amplitude value can be reconstructed by the sampling filter. Further, the accuracy with which the amplitude-continuous signal can be quantized into an amplitude-discrete signal is determined by the accuracy of realizing the multiples of the chosen quantization unit. Notice that the quantization processing can be completed by a digital data compression procedure. Such a procedure consists of encoding and decoding phases. In the encoding phase, an input string of integer-based (I) bits is transformed into a coded string of complex numbers-based (C) bits. The ratio I:C is the compression ratio. In the decoding phase, the compressed string regenerates the original digital data.

The professional sound quality offered by the compact disc digital audio system is in essence obtained by low-noise uniform sampling and waveform quantization, accurate encoding and decoding of the digital audio signals, and, in addition, protection of the encoded digital audio information against disc errors. The CD-A format includes standards for disc specification and signal encoding. The two analog source signals (left and right acoustic audio channels) that originate from the studio or concert hall are converted from Analog to Digital signals (A/D). The clock frecuency at which the two acoustic audio signals are quantized is quartz-crystal-controlled and is equal to 44.1 kHz. This specific sampling frequency is determined by the Phase Alternation Line (PAL) television standard and allows a recorded audio bandwidth of 20 kHz. The uniform amplitude quantization uses 16 bit Pulse Code Modulation (PCM). As an error correction code, the CD-A system employs a doubly encoded Reed-Solomon code called Cross Interleave Reed-Solomon Code (CIRC), which can fully correct burst errors of up to 4000 bits. Such an error is equivalent to a 2.5 mm dropout. The audio samplers are gathered in frames of 12 audio samples each, 6 samples from the left acoustic audio channel and 6 samples from the right channel. To such a frame, parity bytes and control and display bits are added and a specific channel code called Eight to Fourteen Modulation (EFM) applied. In EFM which was especially developed for the CD-A system, a sequence of 8 bits (1 byte) is mapped onto 14 channel bits. A sequence of frames is transferred to the master disc at a channel data rate of 4.32 Mbit/second.

The CD-A standardized format is optically recorded on the surface of a glass disc which is coated with photoresist. The result is the so-called master disc. By galvanic processing, the master disc surface is transferred into a nickel shell. From this "father", stampers are made which are suitable "sons" for replication. By compression or injection molding, the information string contained on the surface of the stamper is transferred in the form of a sequence of about a billion minute pits to a transparent plastic disc. This CD-A which carries a reflective alluminum coating has a diameter of 12 cm, a thickness of 1.2 mm and a track pitch of 1.6 μm. In the CD-A player, the helical track on the disc is optically scanned by an AlGaAs laser of wavelenght 0.8 μm at a constant velocity of about 1.25 m/second. The noise shaper which forms the main component of a recently developed D/A-converter silicon chip in bit-stream VLSI technique allows to minimize the quantization error of PCM by a sequence of 1-bit D/A-conversions.

The Compact Disc Read Only Memory (CD-ROM), based on the CD-A system, records digital data instead of acoustic audio signals and is designed as a nonvolatile, high-speed data storage for computers. Thus the information is not lost when the power is interrupted. The CD-ROM technology is particulary suited for the mass delivery of information systems and data bases that either require or utilize a large amount of computational preprocessing to allow a real-time or interactive respose to be achieved. It uses an identical frame structure as the CD-A system. In a data track the data are divided into addressable blocks of 2352 sequential bytes which contains a CD-A frame and a subcode frame of 98 frames. However, CIRC is not good enough to store data, because interpolation after CIRC cannot be used for digital data. The ECC (Error-Correction Code) based on Reed-Solomon codes over $GF(2^8)$ efficiently transforms long burst errors and multiple short bursts into random errors such that uncorrectable error patterns seldom occur. The laser optical pickup sections of CD players and optical computers disc drives are similar and can be used for CD-A and CD-ROM applications. The access speed of the CD-ROM system is as high as that of a floppy diskette system.

High performance (high rate/high capacity) memory systems are crucial to high resolution image processing. For example, a Synthetic Aperture Radar (SAR) system (see Section 12 *infra*), expected to generate large quantities of data at high rates (up to 300 Mbit/second or greater) can begin operation while storing data on a high rate, high capacity downlink buffer. Presently, the most mature technology for handling high data rates and capacity which is ready for large scale applications are the erasable magneto-optic discs (CD-E), combining the erasability of magnetic storage with the high capacity and permanence of optical storage. A $5\frac{1}{4}$-inch magneto-optic disc, holding 600 Mbytes on both sides, can store up to 1000 times more information than a floppy diskette of the same size.

In the CD-E technology, data are recorded and erased on a thin film of magnetic alloy coatings as gadolinium-terbium-iron, terbium-iron-cobalt, and terbium-iron. The magnetic layer is embedded within a 1.2 mm layer of plastic or glass that protects it from dust, wear, and other problems that plague conventional magnetic recording techniques. The information is stored as a sequence of bits, whose magnetic field is either spin-up (a digital 1) or spin-down (a digital 0).

A blank CD-E has all its magnetic domains pointing spin-down. The important point is that the magnetic field required to flip one domain from spin-down to spin-up —resulting in the coercive force— varies greatly with temperature. At room temperature, however, that field is so high (typically more than 400,000 ampères per meter) that any ordinary magnet is an order of magnitude too weak to change the domain's orientation. Thus also domains will remain spin-down even in the presence of a spin-up bias magnet in the disc drive. However, at about 150°C, the coercive force required to flip a domain falls almost to zero, and the bias is strong enough to record a spin-up state.

To record information, a laser in the electro-optic head with an infrared wavelenght of about 800 nanometers and drawing 8 milliwats heats a spot on the disc 1μm across to 150°C for a few nanoseconds; during that time, the bias magnetic field flips all the domains in the 1μm spot from spin-down to spin-up, recording a bit.

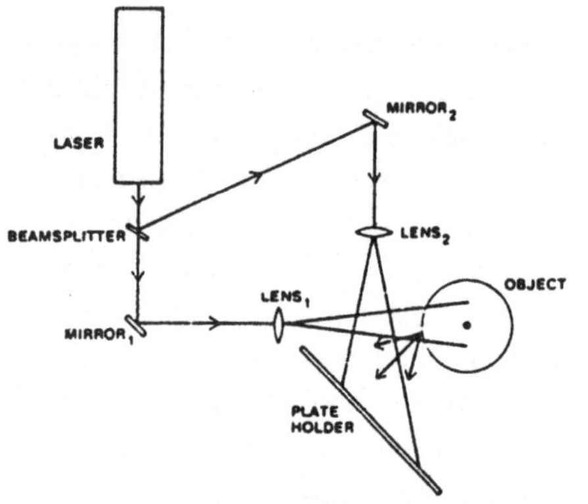

FIGURE 1.

When the laser is turned off, the medium cools inmediately, freezing the bit's domain in the spin-up orientation. During recording the disc rotates and the laser beam is modulated by the data stream, with spin-up bits recorded whenever the beam is on.

To erase information, the bias magnetic field is reversed to spin down. Depending on the system, this is done either by physically inverting the magnet or by reversing the current flow through an electromagnet. With the laser on continuosly, every spin-up bit to be erased flips spin down. New data can be then recorded.

Reading the information takes a low-power beam from the same laser in the electro-optic head. Because of the Kerr magneto-optic effect, the laser beam's plane of polarization will be rotated either clockwise or counterclockwise, depending on whether the magnetic bit is spin-up or spin-down. Optoelectronic circuitry in the electro-optic head senses the polarization, and the bit's orientation is interpreted as a 1 or 0.

To increase reliability, information on a CD-E is written in preformed helical or concentric grooves $1\mu m$ wide and 70 nanometers deep. The electro-optic head senses the groove using light diffraction techniques and keeps the focus laser beam on the center of the groove with servo mechanisms, eliminating physical misalignement of the head, a common problem with magnetic drives.

With the basic modular building block composed of a CD-E and an associated electro-optic head, optical disc buffers can be configured. Capacity per modular element can range from 1–5 Gbits using $5\frac{1}{4}$ inch discs up to 60–80 Gbits using both sides of a 14 inch disc. Data rates can range to 150 Mbit/second using laser diode arrays for each electro-optic head. A combination of twenty-four of the basic building blocks offers terabit capacity and Gbit/second data rates in a single structural configuration.

2.4 Parallel two-dimensional data compression

In those days there was quite a debate raging on whether image processing should be done with optics or computers. One evening, while thinking about the papers that I had heard, it occurred to me that if the opticists had, so to speak, an ace in the hole it was their ability to generate a high resolution, two dimensional Fourier transform using their own version of an FFT computer, namely a lens. Moreover, this "computer" could do it at the speed of light, at extremely high resolution, and do it as quickly for a 10^6-pixel image as for a 100-pixel one.

Henry Stark (1982)

Another consequence of the classical Paley-Wiener theorem outlined in Section 2 *supra* is the quadrature formula

$$\int_{\mathbf{R}} u(t)\,\bar{v}(t)\,dt = \sum_{\mu\in\mathbf{Z}} u(\mu)\,\bar{v}(\mu), \tag{4.1}$$

which holds for all amplitude modulation envelopes u, v in the complex vector space $PW(\mathbf{C})$. The quadrature formula (4.1) has the following physical meaning: The interference pattern of two coherent waves u, v that are mixing synchronously and without phase difference in the holographic plane $\mathbf{R}\oplus\mathbf{R}$ can be compressed by a uniform sequence of coherent pulses. Of course, the shorter the pulses the higher the data compression. In general, however, the two interfering coherent waves are neither synchronous nor without phase difference. If $x\in\mathbf{R}$ denotes the pathway difference and $y\in\mathbf{R}$ the phase difference of two writing laser beams, (4.1) takes the form of a parallel data compression formula:

$$\int_{\mathbf{R}} u(x+t)\,\bar{v}(t)\,e^{2\pi i y t}dt = \sum_{\mu\in\mathbf{Z}} u(x+\mu)\,\bar{v}(\mu)\,e^{2\pi i y \mu}.$$

The left hand side is called the holographic transform $H(u,v;x,y)$ of $u\otimes v$ at the point (x,y) of the holographic plane $\mathbf{R}\oplus\mathbf{R}$ where u, v belong to $PW(\mathbf{C})$. In radar technology, the mapping $(x,y)\mapsto H(u,v;x,y)$ is referred to as the cross-ambiguity function.

The holographic transform $H(u,v;\cdot,\cdot)$ geometrically encodes the interference pattern into the holographic plane $\mathbf{R}\oplus\mathbf{R}$. In optical holography, the practical procedure to simultaneously encode the combined amplitude and phase distribution of the interference pattern of two coherent waves on a light-sensitive recording medium is to divide by a beamsplitter a laser beam into two beams (see Figure 1). The reference beam is sent directly toward the holographic plane $\mathbf{R}\oplus\mathbf{R}$, while the signal beam is aimed at the object being holographed. The signal beam is identical to the coherent reference beam until it strikes the front of the object. As soon as it hits the object it is modulated according to the physical dimensions and characteristics of the object. The wave that is scattered by the object toward the holographic plane $\mathbf{R} \oplus \mathbf{R}$ deviates in amplitude and phase from the virtually unchanged reference wave. Notice that the coherent length of the laser source places an upper limit on the allowable pathway difference between signal and reference beam. The difference is a strong function of the object, carrying information about its surface on a microscopic level. This is precisely what makes holographic image processing such a unique sensing technique. The intensity hologram forms a geometric encoding of the combined amplitude and phase distribution, simultaneously recorded in the holographic plane $\mathbf{R} \oplus \mathbf{R}$. As a recording medium, the emulsion of a photographic film plate, a photorefractive crystal, a Charge-Coupled Device (CCD) detector array, or Charge-Injection Device (CID) camera can be used.

It can be established that the holographic transform $H(u, v; ., .)$ extends to a convolution filter k_f on the complex Hilbert space $L^2(R)$ which is not shift invariant; see formula (5.1) of Section 5 *infra*. For $f \in L^2(\mathbf{R} \oplus \mathbf{R})$ the kernel function k_f takes the form

$$k_f(x, t) = \mathcal{F}^2 f(x - t, t), \tag{4.2}$$

where $\mathcal{F}^2$ denotes the Fourier transform with respect to the second variable. The filter $k_f \in L^2(\mathbf{R} \oplus \mathbf{R})$ can be numerically handed by means of a FFT algorithm in order to produce computer generated holograms (CGHs). Hardware implementations based on VSLI digital FFT processors of the action

$$K_f\, u(x) = \int_{\mathbf{R}} k_f(x, t)\, u(t)\, dt \qquad (x \in \mathbf{R}),$$

are now commercially available. For a cascaded acousto-optic implementation of a programmable real-time synthetic-aperture imaging radar (SAR) processing architecture, see Section 9 *infra*.

For u, v in $L^2(R)$ and $\nu \in \mathbf{R}$, $\nu \neq 0$ the coloured holographic transform is defined by the formula

$$H_\nu(u, v; x, y) := \int_{\mathbf{R}} u(x + t)\, \bar{v}(t)\, e^{2\pi i \nu y t} dt.$$

Thus $H(u, v; \cdot, \cdot) = H_1(u, v; \cdot, \cdot)$. In the case $u = v$, define $H_\nu(u; \cdot, \cdot) := H_\nu(u, u; \cdot, \cdot)$ for $\nu \neq 0$.

2.5 The holographic geometry is sub-Riemannian

Presently, "reading the mind" by multielectrode-arrays is not only technically difficult, but the interpretation of such signals as points in an n-space is theoretically underdeveloped since the underlying (non-Riemannian) CNS geometry is yet unexplored.

András J. Pellionisz (1988)

The three-dimensional real Heisenberg nilpotent Lie group G has as its base manifold M the unipotent matrices

$$\begin{pmatrix} 1 & x & z \\ 0 & 1 & y \\ 0 & 0 & 1 \end{pmatrix} = (x, y, z) = \vec{x},$$

and as its center C the one-dimensional manifold of matrices

$$\begin{pmatrix} 1 & 0 & z \\ 0 & 1 & 0 \\ 0 & 0 & 1 \end{pmatrix} = (0, 0, z).$$

Obviously C is isomorphic to $\mathbf{R}$ and the commutator of any two elements of G belongs to C. Thus the Heisenberg group G is a two-step nilpotent Lie group.

Let TM denote the tangent bundle and T^*M the cotangent bundle of the base manifold M. There exists a unique parabolic bundled form Q on T^*M such that on the diagonal

$$Q_{\vec{x}}(\vec{\xi}, \vec{\xi}) = (\xi - y\zeta)^2 + (\eta + x\zeta)^2$$

holds for all $\vec{\xi} = (\xi, \eta, \zeta)$ in the fibre $T^*_{\vec{x}}M$ of T^*M at the base point $\vec{x} = (x, y, z) \in M$. The annihilator $S_{\vec{x}}$ of the kernel of $Q_{\vec{x}}$ in $T_{\vec{x}}M$ is spanned by $\{(1, 0, -y), (0, 1, x)\}$ and the subbundle S of TM has $S_{\vec{x}}$ as its fibre at the point $\vec{x} \in M$. Define a linear mapping

$$g(x) : T^*_{\vec{x}}M \longrightarrow T_{\vec{x}}M$$

by the prescription

$$Q_{\vec{x}}(Y, g(\vec{x})\vec{\xi}) = \langle Y, \vec{\xi} \rangle \qquad (Y \in S_{\vec{x}}).$$

Then g defines a sub-Riemannian metric on S. There is a notion of geodesic for the associated sub-Riemannian geometry (see STRICHARTZ [25]): Define the Hamiltonian function

$$\mathcal{H}(\vec{x}, \vec{\xi}) = \tfrac{1}{2}\langle g(\vec{x})\vec{\xi}, \vec{\xi} \rangle$$

on T^*M. Then the smooth curves $\vec{x}(t)$ in M are called geodesics if there exists a cotangent lift $\vec{\xi}(t) \in T_{\vec{x}(t)}M$ such that $g(\vec{x})\vec{\xi} = (\vec{x})$ and the Hamilton-Jacobi equations

$$\dot{\vec{x}} = \nabla_{\vec{\xi}}\mathcal{H}, \qquad \dot{\vec{\xi}} = \nabla_{\vec{x}}\mathcal{H}$$

are satisfied in an appropiate open parameter interval of the real line $\mathbf{R}$. It turns out that the geodesics in M are the Heisenberg helices. Further, in the Heisenberg sub-Riemannian geometry of M, small neighborhoods are similar to large neighborhoods, so nothing is gained by working locally on the Heisenberg nilpotent Lie group G in contrast with Riemannian geometry.

The Heisenberg group G is at the basis of quantum mechanics and signal theory; see the author's monograph [21]. For each real number $\nu \neq 0$ it admits a unique (up to isomorphy) irreducible unitary linear representation U_ν that subduces on $\mathbf{C}$ the character

$$z \mapsto e^{2\pi i \nu z}.$$

U_ν is called the linear Schrödinger representation of G. Its projection along the center C gives rise to the fundamental formula for the coloured holographic transform

$$H_\nu(u, v; x, y) = \langle U_\nu(x, y, 0)u|v\rangle \tag{5.1}$$

where $\langle\cdot|\cdot\rangle$ denotes the standard scalar product of the complex Hilbert space $L^2(\mathbf{R})$.

2.6 Holographic reciprocity and coupling

Die Reziprozität zwischen Signal- und Referenzwelle ist charakteristisch für die Holografie. Sie bildet eine wichtige Grundlage für ihre Anwendung.

Sigurd Kusch (1987)

The identity (5.1) allows to transfer the properties of the linear Schrödinger representation U_ν of G to the coloured holographic transform H_ν. Among these properties the most important are the identification by duality of the (flat) Kirillov coadjoint orbit associated with U_ν to the projection G/C of the Heisenberg group G along its center C which is the holographic plane $\mathbf{R} \oplus \mathbf{R}$, the square integrability uf U_ν modulo C, and the disjointness of U_ν and $U_{\nu'}$ for $\nu \neq \nu'$. In particular we get the identities

$$\langle H(u', v'; \cdot, \cdot)|H(u, v; \cdot, \cdot)\rangle = \langle u' \otimes v|u \otimes v'\rangle, \tag{6.1}$$
$$\langle H_\nu(u', v', \cdot, \cdot)|H_{\nu'}(u, v; \cdot, \cdot)\rangle = 0 \quad (\nu \neq \nu'), \tag{6.2}$$

for all $u', v', u, v \in L^2(\mathbf{R})$.

The Four-Wave Mixing (FWM) formula (6.1) which can be traced back to the work of V. Bargmann is a consequence of Schur's lemma. It is at the basis of the filter extension k_f of the holographic transform $H(u, v, \cdot, \cdot)$ mentioned in Section 2 *supra*. Indeed, according to the Stone-von Neumann-Segal theorem the mapping $f \mapsto k_f$ extends to an isometric isomorphism

$$f \mapsto U_1(f) = K_f$$

of $L^2(\mathbf{R} \oplus \mathbf{R})$ onto the complex Hilbert space of Hilbert-Schmidt operators acting on $L^2(\mathbf{R})$. From this Payley-Wiener type theorem we infer the holographic reciprocity principle which dominates the geometric decoding of holograms:

If an intensity hologram recorded in the holographic plane $\mathbf{R} \oplus \mathbf{R}$ is illuminated by the coherent reference wave then the signal wave is diffracted. Conversely, illumination by the signal wave generates the reference wave.

Therefore holographic decoding or wavefront reconstruction takes place upon scanning the magnified hologram with a suitably scaled coherent reference beam.

In the paper [23], the FWM formula (6.1) has been used to compute the coupling coefficients of laser modes guided by optical fibers. Finally, formula (6.2) is at the basis of the C^3-laser (cleaved-coupled-cavity laser; *cf.* TSANG *et al.* [26]), and the LECOS network of the fly-by-light system.

2.7 Elementary holograms and complete bipartite graphs

Welche Möglichkeiten haben wir überhaupt, Strukturen oder Vorgaänge in ihnen zu verstehen. Ein beliebtes und oft auch erfolgreiches Vorgehen ist kleinere Einzelteile zu zerlegen.

Hermann Haken (1981)

The fundamental formula (5.1) gives rise to a geometric characterization of the transverse Gauss-Hermite modes of a laser beam. Indeed $u \in L^2(\mathbf{R})$ is called radially isotropic if $H(u, \cdot, \cdot)$ is a radial function on the holographic plane $\mathbf{R} \oplus \mathbf{R}$, i.e., invariant under the natural action of SO(2,**R**). It can be established that $u \in L^2(\mathbf{R})$ is radially isotropic if and only if u is up to a complex constant a Hermite function (or harmonic oscillator wave function) H_n of arbitrary degree $n \geq 0$. The proof depends upon the diamond solvable Lie group, the semi-direct product of the Heisenberg group G with the one-dimensional compact torus group T.

The elementary holograms are defined as the images of $H_m \otimes H_n$ under the holographic transform H. The explicit formula for the elementary holograms reads as follows:

$$H(H_m, H_n; w) = \sqrt{\frac{n!}{m!}} (\sqrt{\pi} w)^{m-n} L_n{}^{(m-n)}(\pi |w|^2) \qquad (7.1)$$

where $w = x + iy$ is an arbitrary point of the complexified holographic plane $\mathbf{R} \oplus \mathbf{R}$, $L_n^{(p)}(w) = e^{-w/2} l_n^{(p)}(w)$ denotes the Laguerre function of degree n and parameter p, and $m \geq n \geq 0$. The elementary holograms can be computed by the three-term recurrence relation of the Laguerre polynomials $l_n^{(p)}$. They can also be expressed in terms of the Charlier-Poisson polynomials (see [28]) and the generalized Bateman functions (see YAACOB [29]). In Section 13 *infra* the elementary holograms will be expressed by the matching polynomials $\Phi_{K_{m,n}}$ of the complete bipartite graphs $K_{m,n}$ with $m+n$ vertices. Indeed we have for $w \in \mathbf{R}$:

$$\Phi_{K_{m,n}}(w) = (-1)^n n!\, w^{m-n}\, l_n^{(m-n)}(|w|^2) \qquad (m \geq n), \tag{7.2}$$

and the coefficients $\Phi_{K_{m,n}}$ count the number of ways in which disjoint edges can be selected in $K_{m,n}$.

It should be observed that the elementary hologram (7.1) can be lifted from the projection G/C along the center C to the whole Heisenberg Lie group G. The lifted elementary holograms form the eigenfunctions of the sub-Laplacian L of the Heisenberg group G. Notice that sub-Riemannian geometry is to L what Riemannian geometry is to the Laplacian.

The hypoelliptic linear differential operator L forms the Hamiltonian of a spinless particle in a constant external magnetic field. Since the spin does not influence the wave function u of an electron in a constant magnetic field (*cf.* VON KLITZING [15]), formula (7.1) also describes the elementary electron holograms. For the topic of electron holography, see the papers by TONOMURA [27] and LICHTE [17].

Recently, J.E. Trebes and his collaborators at Lawrence Livermore National Laboratory (LLNL) succeeded in generating the first X-ray holograms.

2.8 Holographic invariants and linear optical phase conjugation

The extent of the debt phase conjugation owes to holography is not clear. Nor it is even clear just where true holography stops and phase conjugation begins.

Emmet N. Leith (1985)

A mapping $h : \mathbf{R} \oplus \mathbf{R} \longrightarrow \mathbf{R} \oplus \mathbf{R}$ is called an orientation-preserving holographic invariant if the covariant identity of the metaplectic representation

$$H(u; x, y) = H(u_h; h(x, y)), \tag{8.1}$$

holds for all points (x, y) of the holographic plane $\mathbf{R} \oplus \mathbf{R}$ and all functions $u \in L^2(\mathbf{R})$ in such a way that the assignement $u \mapsto u_h$ defines a unitary operator in the complex Hilbert space $L^2(\mathbf{R})$. It can be established that h is an orientation-preserving holographic invariant mapping if and only if $h \in SL(2, \mathbf{R})$, *i.e.*,

$$h = \begin{pmatrix} a & b \\ c & d \end{pmatrix}, \qquad \det h = 1$$

where a, b, c, d are real entries. In particular, h is an invertible linear mapping of the holographic plane $\mathbf{R} \oplus \mathbf{R}$ onto itself.

Notice that the mapping $h \in SL(2, \mathbf{R})$ induces a time-transformed replica u_h of $u \in L^2(\mathbf{R})$. For example, if h is the symplectic matrix

$$j = \begin{pmatrix} 0 & 1 \\ -1 & 0 \end{pmatrix}, \tag{8.2}$$

which forms the infinitesimal generator of the orientation-preserving rotations about the origin of the holographic plane $\mathbf{R} \oplus \mathbf{R}$, then the image u_j of $u \in L^2(\mathbf{R})$ under the action of $j \in SL(2, \mathbf{R})$ is the Fourier transform $\mathcal{F}u$ of u:

$$u_j = \mathcal{F}u \qquad (u \in L^2(\mathbf{R})).$$

Thus, the rotation (8.2) about the origin of the holographic plane $\mathbf{R} \oplus \mathbf{R}$ through the directed angle of 90° yields the frequency spectrum of the amplitude modulation envelope u. The identity

$$u_{j^2}(t) = \mathcal{F} \circ \mathcal{F}\, u(t) = u(-t) = u^-(t) \qquad (t \in \mathbf{R}),$$

which forms the main step in the proof of the Fourier inversion formula for $u \in L^2(\mathbf{R})$, shows that the rotation j^2 about the origin of the holographic plane $\mathbf{R} \oplus \mathbf{R}$ through the directed angle of 180° generates a time reversal $u \mapsto u^-$. Thus j^2 performs a linear phase conjugation which transforms the pseudoscopic image of a hologram into the orthoscopic image.

2.9 Cascaded acousto-optic real-time kernel implementation

> ***Optical signal processing technology is so well matched to the SAR data processing requirements that optical processors still acount for a major portion of the SAR images produced, despite the tremendous advances made in digital signal processing technology over the last thirty years.***
>
> Demetri Psaltis and Michael Haney (1987)

One of the most successful applications of microwave holographic image processing technology has been in the formation of high resolution terrain images from data collected by a Synthetic-Aperture imaging Radar (SAR) system. The principal advantage of SAR is its ability to form such images of shrouded terrains at relatively long wavelengths with small antennas at all days, *i.e.*, day or night. The disadvantage is that the signals collected by a SAR system are not already focused as is the case in real-aperture imaging systems, and require extensive processing to form the SAR image from the received radar signal.

A SAR imager that operates in real time must have the ability to adapt rapidly to dynamic parameter changes of the radar-target geometry to generate a well-focused image continuosly. Figure 2 displays a programmable

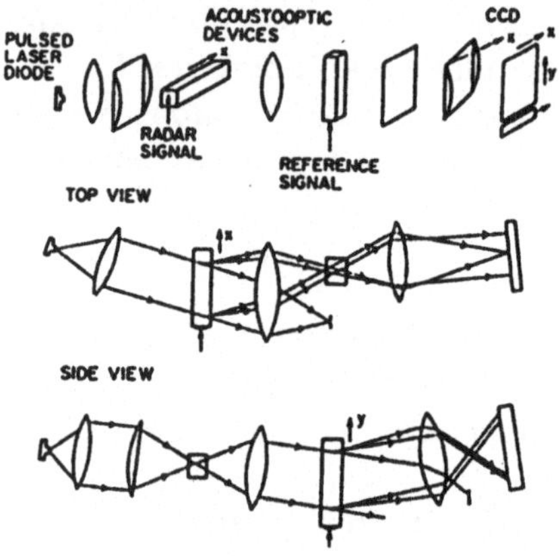

FIGURE 2.

real-time SAR processing architecture which optically computes from $f \in L^2(\mathbf{R}\oplus\mathbf{R})$ the kernel function (4.2) of the Hilbert-Schmidt operator $U_1(f) = K_f$ acting on $L^2(\mathbf{R})$ (see Section 4 *supra*) by an application of the rotation about the origin of the holographic plane through the directed angle of 90°. In this cascaded implementation the interfering coherent beams pass through the same anamorphic optical system and the Acousto-Optic Device (AOD) containing the radar signal (see PSALTIS [19]). The two coherent waves interfere at the holographic plane $\mathbf{R} \oplus \mathbf{R}$ where the Charge-Coupled Device (CCD) detector arrays are located to provide high resolution output.

The motivation for using AODs comes from the fact that they are well developed at least by comparison with other Spatial Light Modulators (SLMs), and suitable for the practical implementation of acousto-optic signal processors. An AOD consists of a piezo-electric transducer bonded to a transparent medium capable of supporting sound propagation. An electrical signal input to the transducer is converted into an ultrasonic wave that carries both the amplitude and the phase of the electrical signal. The traveling ultrasonic wave manifests itself as a density wave which through the photoelastic effect results in an index of refraction variation in the acousto-optic material. This index variation, or grid, carries the combined amplitude and phase information of the ultrasonic wave and thus of the original electrical input. The amplitude and phase of the resulting light diffracted by the grid is thus proportional to that of the original electrical input.

The CCD is an analog device that does not require A/D-coversion and operates as a shift register in which sampled values of the analog input signal are stored in the form of charges on an array of capacitors. Switches between capacitors transfer the charge from one capacitor to the next, following a command from an applied voltage pulse. In view of the uniform sampling theorem (Section 2 *supra*), the maximum bandwidth that can be processed without distortion is given by half of the clock frequency. For conventional CCDs, the highest clock frequency consistent with good transfer efficiency is around 10 MHz. This implies that bandwidths of up to 5 MHz can be comfortably processed. Once the input analog signal has been sampled and transferred into the CCD, it consists of discrete charge packets which migrate out of the pixel collectors during readout erasing the image stored on the sensor. The recently developed highly sophisticated Charge Injection Device (CID) cameras, however, accomplish image readout by using a charge transfer technique that reads displacement current when collected charge packets are shifted between capacitors within a single addressed pixel. Charge does not transfer from site to site in the arrays as with CCDs. This displacement current is proportional to the stored signal charge, so the value can be fed to the outside as a video signal that accurately reflects photon energy collected at each pixel. The readout mechanism is non-destructive because the charge remains intact at the pixels after the signal level has been determined. Switching the row and column electrodes to ground empties the pixel collectors by releasing or injecting the charge packets into the collector in the substrate, clearance array in preparation for new image integration. This fundamental principle of operation makes CID different from other imaging technologies. Because there are virtually no opaque areas between pixels where photon-generated charge can be lost, fill factors generally exceed 90% . This contiguous pixel structure provides the capability for extremely accurate edge detection using sub-pixel interpolation processing techniques.

Charge transfer devices like CCDs and CIDs offer a number of advantages over conventional Cathode Ray Tubes (CRTs) such as high sensitivity and wide spectral and dynamic range. Combined with acousto-optic processors they are able to replace the photographic film in the classical SAR processing architecture (see Section 12 *infra*) and to adapt this architecture so that it operates in real time (see PSALTIS [19]).

2.10 Classification of pixel mappings and holographic interferometry

Während die Fünfecksymmetrie in der organischen Welt häufig vorkommt, findet sie sich nicht bei den vollkommensten symmetrischen Schöpfungen der anorganischen Natur, bei den Kristallen. Dort sind keine anderen Drehsymmetrien möglich als die der Ordnung 2, 3, 4 und 6.

Hermann Weyl (1955)

In analogy to the uniform sampling theorem of digital signal processing (see Section 2 *supra*), the quadratic lattice $\mathbb{Z} \oplus \mathbb{Z}$ of Gaussian integers $Z[i]$

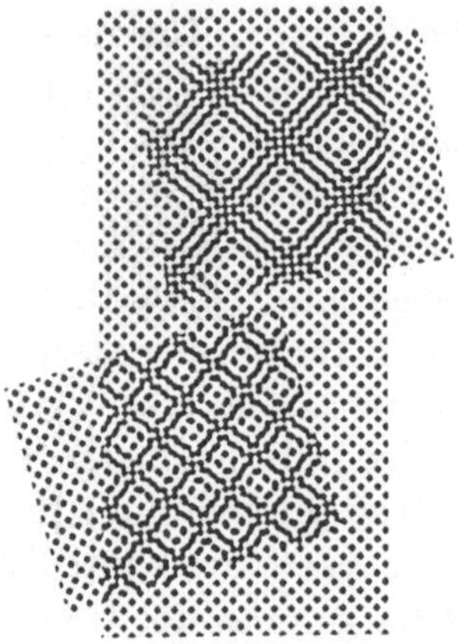

FIGURE 3.

can be embeded into the (complexified) holographic plane $\mathbf{R} \oplus \mathbf{R}$ and the holographic identity

$$\sum_{(\mu,\nu)\in\mathbf{Z}\oplus\mathbf{Z}} H(u;\mu,\nu)\,\overline{H}(v;\mu,\nu) = \sum_{(\mu,\nu)\in\mathbf{Z}\oplus\mathbf{Z}} |H(u,v;\mu,\nu)|^2 \qquad (10.1)$$

arises.

An orientation-preserving holographic invariant $h \in SL(2,\mathbf{R})$ is called a pixel mapping if

i) h preserves the lattice $\mathbf{Z}[i]$ in $\mathbf{C}$, and

ii)u_h is radially isotropic for all wave functions $u \in L^2(\mathbf{R})$.

For a pixel mapping h we get from (10.1) the following identity for elementary holograms

$$\sum_{(\mu,\nu)\in\mathbf{Z}\oplus\mathbf{Z}} H(H_m;h(\mu,\nu))\overline{H}(H_n;h(\mu,\nu)) = \sum_{(\mu,\nu)\in\mathbf{Z}\oplus\mathbf{Z}} |H(H_m,H_n;h(\mu,\nu))|^2 \qquad (10.2)$$

which is valid for all degrees $m \geq n \geq 0$. For a discussion of the identity (10.2) in the framework of the Moyal calculus, see the paper [28].

The unimodular Möbius transformation associated with h has to be elliptic. Therefore, if h is not the identity of the holographic plane $\mathbf{R}\oplus\mathbf{R}$, the trace h has to satisfy the condition

$$-2 \leq \operatorname{tr} h \leq +2.$$

Consequently the rotations of order

$$k \in \{1,2,3,4,6\}$$

in the holographic plane $\mathbf{R} \oplus \mathbf{R}$ are the only pixel mappings. This classification implies that there are five Euclidean orientable 3-orbifolds of planar

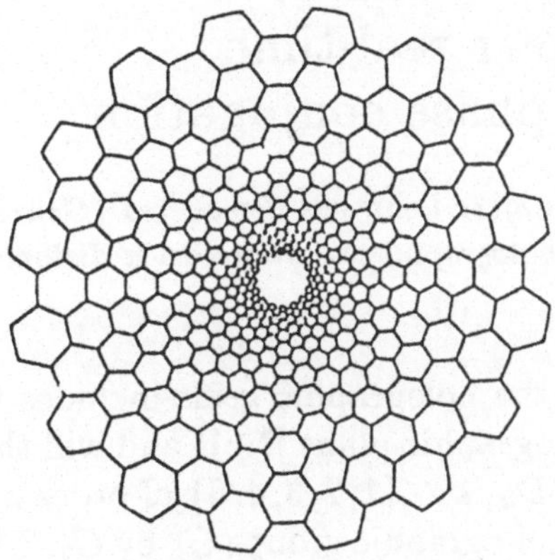

FIGURE 4.

holographic grids which correspond to the orientation preserving cristallographic groups $C_k, k \in \{1, 2, 3, 4, 6\}$ (see the paper [22]).

Most engineering applications of optical holography make use of the ability to record the superposition of two slightly different images and display the minute differences between them. This technique is called holographic interferometry. A powerful feature if the holographic interferometry technique is that information is obtained from the entire test object rather than just at a single point. Figure 3 illustrates the superposition of two elements of the 3-orbifold of quadratic holographic grids ($k = 4$). Notice that the Moiré patterns created by holographic interferometry show a magnified scale. In this way holographic interferometry allows a user to visualize by a CCD detector array and quantify minute deformations in a test object. Interferometric contour lines provide the precise dimensional data for computer-aided design and finite element analysis.

The existence of holographic grids has been experimentally established by Pál Greguss[1] using a Majoros-type toroidal lens (*cf.* GREGUSS [9]). In neurophysiology, planar holographic grids occur in the retino-cortical channel which transmits visual information to the cerebral cortex in Pulse Code Modulation (PCM) form. Specifically, we mention the hexagonal grid in the extra-foveal retina of primates (see Figure 4), the hexagonal presynaptic vesicular grids of the synaptic clefts, the hexagonal excitation patterns in the area 17 of the visual cortex, and the hexagonal symmetry of the patterns underlying visual hyperacuity. Thus the holographic hypothesis fits the tendency toward geometrization in neurophysiology.

The case $k = 4$ allows to establish via j the uniform sampling theorem from the holographic point of view and to consider the CD-A, CD-ROM, and CD-E as one-dimensional holograms (see Section 3 *supra*).

---

[1] Applied Biophysics Laboratory, Technical University, Budapest.

2.11 Non-linear real-time optical phase conjugation

A short time after light was discovered (Genesis 1:3), Kogelnik demonstrated that holography can undistort a distorted image.

Jack Feinberg (1985)

The full symmetry of the holographic grids includes the reflection $w \mapsto \overline{w}$ of the complexified holographic plane $\mathbf{R}\oplus\mathbf{R}$ and will therefore be described by the dihedral groups D_k, $k \in \{1,2,3,4,6\}$ of order $2k$ which are semi-direct products of the crystallographic groups C_k by C_2. The Poncaré extension to $\mathbf{R}^3$ of the aformentioned reflection $w \mapsto \overline{w}$ gives rise to a time reversal $u \mapsto u^-$. It transforms the hologram formation geometry using a photorefractive crystal into the readout geometry and therefore performs by degenerate FWM real-time holographic reciprocity a phase-conjugate replica of the signal wave.

Particularly in the experimental field, great progress has been made over the past several years to find photorefractive crystals such as $BaTiO_3$ having a light sensitivity comparable with that of optical film. However, $BaTiO_3$ is too slow as far as the response time is concerned. The Bismuth Silicon Oxide (BSO) crystal ($Bi_{12}Si_{20}$) is an attractive electro-optical material for dynamic holography and real-time phase-conjugate wavefront reconstruction. From the holographic grid point of view it is clear that the photorefractive effect in such a phase-conjugator cannot be spatially local.

It is possible to make a phase conjugator with optical gain, so that a weak input signal is both phase-conjugated and amplified. This has the following interesting consequence: if a reflective surface is placed facing the phase-conjugator, then self-oscillation between the mirror and the phase-conjugator will occur, and a beam will be formed between them. If the mirror is moved, the light beam will automatically track it.

For a survey of optical phase conjugation and its applications, see the papers by FEINBERG [6] and YARIV [30].

2.12 The classical SAR processing architecture

Such optical systems are certainly among the strangest ever devised. They have little versatility, being useful for only one purpose —they carry out this compensation process. They do it, however, remarkably well, in an exceedingly cost-effective way.

Emmet N. Leith (1978)

In Synthetic-Aperture Radar (SAR) systems such as Side-Looking Airborne Radar (SLAR) and Shuttle Imaging Radar (SIR-A, SIR-B), the transmitting and receiving antennas are both mounted on the same platform, and usually one antenna is used for both transmission and reception. A periodic pulse train is transmitted as the platform moves with respect to the ground being mapped, and the sequence of radar echoes from the target scene is received by the radar at locations along the platform's orbit. The received signals are stored until all of the radar returns are collected. The sequence of returns are then processed to generate the image. The coordinates of the holographic plane $\mathbf{R} \oplus \mathbf{R}$ in which the radar image is formed are the range x and the azimuth y. Range is in the direction perpendicular to the direction of the flight of the platform, and the azimuth direction is parallel to the flight.

As the platform travels past each point scattered, the return signal is upshifted in frequency as the scatterer is approached, has a zero Doppler shift at the point where the point scatterer is directly to the side of the radar, and has a negative frequency shift as the platform moves away. The classical SAR processing architecture (see ELACHI *et al.* [5]) is based upon an implementation of the Doppler frequency shift and a scaling factor $p \in \mathbf{R}$ into the holographic transform H. The Doppler effect performs a symplectic transformation of the holographic plane $\mathbf{R} \oplus \mathbf{R}$ which is thought of as the projection G/C of the Heisenberg group G along its center C. The transformation is induced by the matrix

$$h_1 = \begin{pmatrix} 1 & q \\ 0 & 1 \end{pmatrix}$$

where $q \in \mathbf{R}$ depends upon the parameters, namely the constant velocity c_p of the airborne or spaceborne microwave radar which illuminates the target from different perspectives of the platform, the radar wavelenght λ, and the minimum range distance r. Indeed,

$$q = \frac{c_p^2}{2\lambda r}.$$

If $q \neq 0$, it is not difficult to establish that u_{h_1} is the convolution of $u \in L^2(\mathbf{R})$ with a chirp of chirp rate q^{-1}:

$$\mathrm{chirp}_{q^{-1}}(t) = e^{-\pi i t^2/q} \qquad (t \in \mathbf{R}).$$

The Fourier transform of u_{h_1} is the product of the Fourier transform of u, the chirp of chirp rate $-q$, and the term including the Maslov index:

$$\mathcal{F} u_{h_1}(t) = \mathcal{F} u(t)\, \mathrm{chirp}_{-q}(t)\, e^{-\pi i/4\, \mathrm{sign}(q)} \qquad (t \in \mathbf{R}).$$

Consequently, a phase jump of $\pi/2$ occurs at the instant the target is directly to the side of the platform. The astigmatism generated by the Doppler frequency shift can be optically corrected by an anamorphic system including cylindrical lenses between the data film and the image film; see the paper by LEITH [16]. Their optical focal lenght is given by

$$f_r = \frac{c_F^2 \lambda r}{2\lambda_L c_p^2}$$

where c_p denotes the film transport velocity and λ_L the wavelenght of the reading laser beam. The range dependence of f_r can be compensated for by using a tilted cylindrical lens (see Figure 5).

The scaling matrix including the entry $p \in \mathbf{R}$ is given by

$$h_2 = \begin{pmatrix} p & 0 \\ 0 & p^{-1} \end{pmatrix} \qquad (p \neq 0).$$

Notice that the rotation j [see (8.2)] and the matrices of type h_1 and h_2 generate the symplectic group $SL(2, \mathbf{R})$ and that the resolution achievable by the SAR processing is independent of the range r to the target and the velocity c_p of the moving platform. This interesting result reflects the fact that the holographic geometry is sub-Riemannian and therefore of a non-local character (see Section 5 *supra*).

For a survey of SAR, the reader is referred to the monographs by HARGER [10] and FITCH [7].

2.13 Neural computer architecture for pattern recognition

The horizontal cells in many species are connected to each other by gap junctions to form an electrically continuous network in which signals propagate by electronic spread. The voltage at every point in the network thus represents a spatially weighted average of the photoreceptors input. The farther away an input is from a point in the network, the less weight it is given. The horizontal cells are usually modeled as passive cables, in which the weighting function decreases exponentially with distance.

Carver A. Mead and Misha A. Mahowald (1988)

A comparison of the identities (7.1) and (7.2) in Section 7 *supra* yields for the elementary holograms the equation

$$H(H_m, H_n; w) = \frac{(-1)^n}{\sqrt{m!n!}} e^{-\pi|w|^2/2} \Phi_{K_{m,n}}(\sqrt{\pi} w) \qquad (m \geq n),$$

where the matching polynomial in the right hand side is given by

$$\Phi_{K_{m,n}}(\sqrt{\pi} w) = \sum_{0 \leq l \leq [(m+n)/2]} (-1)^l p(K_{m,n}, l) (\sqrt{\pi} w)^{m+n-2l}$$

and $w = x + iy$ is an element of the complexified holographic plane $\mathbf{R} \oplus \mathbf{R}$. The polynomial coefficients involved are given by $p(K_{m,n}, 0) = 1$ and by the

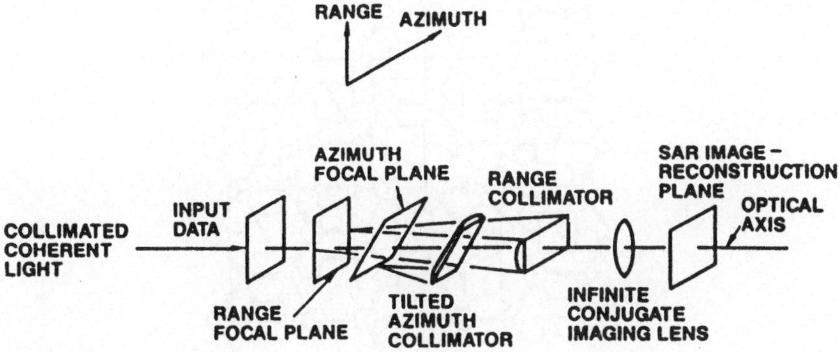

FIGURE 5.

number $p((K_{m,n}, l)$ of ways in which $l \geq 1$ disjoint edges can be selected in the complete bipartite subgraph $K_{m,n}$ with $m+n$ vertices of the holographic grid. For the matching polynomials, see the papers by HEILMANN AND LIEB [11], GODSIL AND GUTMAN [8], and HOSOYA [13].

For instance, in the case $m = 3, n = 3$ the coefficients are $p(K_{3,3}, 0) = 1$, $p(K_{3,3}, 1) = 9$, $p(K_{3,3}, 2) = 18$, $p(K_{3,3}, 3) = 6$, and in the case $m = 4$, $n = 2$ we get the coefficients $p(K_{4,2}, 1) = 8$, $p(K_{4,2}, 2) = 12$. The figures below illustrate the associated complete bipartite graphs $K_{3,3}$ and $K_{4,2}$ with 6 vertices.

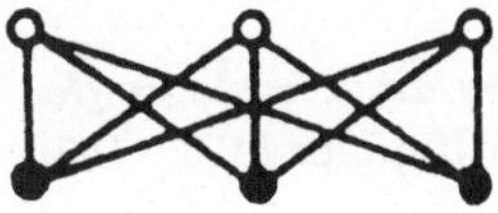

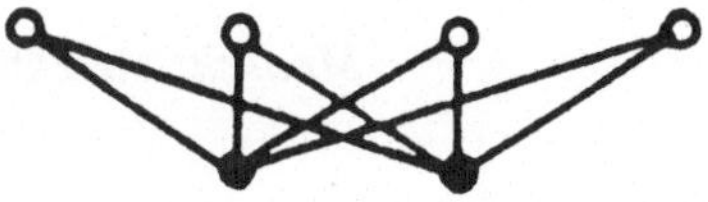

If the holographic grid is realized by a neural network, the bipartition of the complete subgraph $K_{m,n}$ $(m \geq n)$ in the holographic plane $\mathbf{R} \oplus \mathbf{R}$ is given by the firing and non-firing binary formal neurons. The coefficients of the matching polynomial $\Phi_{K_{m,n}}$ together with the exponentially decreasing factor provide the weights of the synaptic interconnections in a hierarchical manner. Recently, C. Mead and his collaborators at CALTECH succeeded in the hardware implementation of a hexagonal neural network (k=6) for image processing. The results of Hubel and Wiesel obtained from their analysis of the mammalian visual cortex underscore the importance of oriented-edge detection. Mead's hardware design to perform orientation pre-processing meets the needs for a feature-based pattern recognition system and is inspired by the structure of animal visual systems. A silicon VSLI retina constructed in standard Complementary Metal-Oxide Semiconductor (CMOS) technique contains an array of photoreceptors and additional hardware to compute by pixel-wise local operation three analog values that encode the magnitude and direction of the intensity gradient at

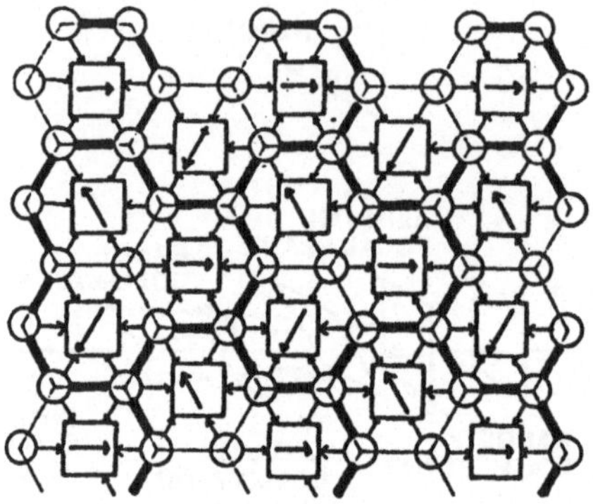

FIGURE 6.

each point of the hexagonal holographic grid (see Figure 6). The simulated self-organized learning network produces an edge-enhanced image, which then passes through an additional stage, located outside the array, that computes the orientation. The outputs of this stage provide an enriched feature set to the next level of the artifitial pattern recognition system. For details, the reader is referred to the papers by MEAD AND MAHOWALD [18] and ALLEN *et al.* [1].

2.14 Conclusions

Mensch, streckh deine Vernunft hieher, diese dinge zu begreiffen!
Johannes Keppler (1571-1630)

Holographic image processing is a coherent sensing technique that can encode and decode high resolution images. The geometric encoding of the two-wave mixing $u \otimes v$ by the combined amplitude and phase distribution in an interference pattern recorded in the holographic plane $\mathbf{R}\oplus\mathbf{R}$ is simultaneously performed by means of the holographic transform $H(u, v; \cdot, \cdot)$. The geometric decoding of the interference pattern is based in the holographic reciprocity principle.

The quantum mechanical approach to signal geometry is based on harmonic analysis on the real Heisenberg nilpotent Lie group, with the Stone-von Neumann-Segal theorem as its central result (see the monograph [21]). It allows us to deal with optical holography including linear and non linear optical phase conjugation, and electron holography in a unified way. Metaplectic harmonic analysis is the basis of transverse tomographic methods in high resolution imaging radar, for instance synthetic-aperture imaging radar like side-looking airborne SAR and space shuttle SAR, and leads to a neural computer architecture for pattern recognition.

2.15 Acknowledgements

The author wishes to thank Professors Dr.Dr.h.c.mult. Hermann Haken (Institute of Theoretical Physics and Synergetics, University of Stuttgart), Dr. José M. Gracia-Bondía, and Dr. Joseph C. Várilly (Escuela de Matemática, Universidad de Costa Rica) for valuable discussions.

2.16 References

[1] T. Allen, C. Mead, F. Faggin, and G. Gribble, Orientation-selective VLSI retina. In: *Visual Communications and Image Processing '88*, T. Russell Hsing, Editor, Proc. SPIE 1001, 1040–1046 (1988).

[2] M.F. Barnsley, J.H. Elton, and D.P. Hardin, Recurrent iterated function systems, *Constr. Approx.* **5**, 3–31 (1989).

[3] P.L. Butzer. W. Splettstößer, and R.L. Stens, The sampling theorem and linear prediction in signal analysis, *Jber. d. Dt. Math.-Verein.* **90**, 1–70 (1988).

[4] F.C. Crow, The aliasing problem in computer-generated shaded images, *Comm. ACM* **20**, 799–805 (1977).

[5] C. Elachi, T. Bicknell, R.L. Jordan, and C. Wu, Spaceborne synthetic-aperture imaging radars: applications, techniques and technology, *Proc. IEEE* **20**, 1174–1209 (1982).

[6] J. Feinberg, Applications of real-time holography. In: *Holography*, Lloyd Huff, Editor, Proc. SPIE 532, 119–135 (1985).

[7] J.P. Fitch, *Synthetic aperture radar* (Springer-Verlag, New York, 1988).

[8] C.D. Godsil and I. Gutman, On the theory of the matching polynomial, *J. Graph Theory* **5**, 137–144 (1981).

[9] P. Greguss, On easy-to-handle holocameras for endoscopic applications. In: *Optics in Biomedical Sciences*, G. von Bally and P. Greguss, Editors, (Springer-Verlag, Heidelberg, 1982), pp. 96-99.

[10] R.O. Harger, *Synthetic Aperture Radar Systems* (Academic Press, New York, 1970).

[11] O.J. Heilmann and E. H. Lieb, Theory of monomer-dimer systems, *Commun. Math. Phys.* **25**, 190–232 (1972).

[12] J.R. Higgins, Five short stories about the cardinal series, *Bull. Amer. Math. Soc.* **12**, 45–89 (1985).

[13] H. Hosoya, Matching and symetry of graphs, *Comp. and Maths. with Appls.* **12B**, 271–290 (1986).

[14] A.J. Jerri, The Shannon sampling theorem —its various extensions and applications: a tutorial review, *Proc. IEEE* **65**, 1565–1596 (1977).

[15] K. von Klitzing, The quantized Hall effect, *Rev. Mod. Phys.* **58***(3)*, 519–531 (1986).

[16] E.N. Leith, Synthetic aperture radar. In: *Optical Data Processing*, D. Casasent, Editor (Springer-Verlag, New York 1978), pp. 89-117.

[17] H. Lichte, Electron holography approaching atomic resolution, *Ultramicroscopy* **20**, 293–304 (1986).

[18] C.A. Mead and M.A. Mahowald, A silicon model of early visual processing, *Neural Networks* **1**, 91–97 (1988).

[19] D. Psaltis, Two-dimensional optical processing using one-dimensional input devices, *Proc. IEEE* **72**, 962–974 (1984).

[20] W. Schempp, Gruppentheoretische Aspekte der Signalübertragung und der kardinalen Interpolationssplines I, *Math. Meth. Appl. Sci.* **5**, 195–215 (1983).

[21] W. Schempp, *Harmonic Analysis on the Heisenberg Nilpotent Lie Group, with Applications to Signal Theory* (Longman Scientific and Technical, Harlow, Essex 1986).

[22] W. Schempp, Elementary holographs and 3-orbifolds, *C. R. Math. Rep. Acad. Sci. Canada* **10***(3)*, 155–160 (1988).

[23] W. Schempp, Extensions of the Heisenberg group and coaxial coupling of transverse eigenmodes, *Rocky Mountain J. Math.* **19** (1988).

[24] L. Schwartz, Sous-espaces hilbertiens d'espaces vectoriels topologiques et noyaux associés (noyaux reproduisants), *J. Analyse Math.* **13**, 115–256 (1964).

[25] R.S. Strichartz, Sub-Riemannian geometry, *J. Differential Geometry* **24**, 221–263 (1986).

[26] W.T. Tsang, N.A. Olsson, and R.A. Logan, High-speed direct single-frequency modulation with large tuning rate and frequency excursion in cleaved-coupled-cavity semiconductor lasers, *Appl. Phys. Lett.* **42***(8)* 639–669 (1987).

[27] A. Tonomura, Applications of electron holography, *Rev. Mod. Phys.* **59***(3)*, 639–669 (1987).

[28] J.C. Várilly, J.M. García-Bondía, and W. Schempp, The Moyal representation of quantum mechanics and special function theory (to appear).

[29] K.B. Yaacob, The hydrogen atom of Bateman functions, *Sing. J. Phys.* **5**, 77–93 (1988).

[30] A. Yariv, Compensation for optical propagation distortion through phase adaption by nonlinear techniques (PANT). In: *Adaptive Optics and Short Wavelenght Sources*, S.F. Jacobs, M. Sargent III, and M.O. Scully, Editors (Addison Wesley, Don Mills, Ontario, 1978), pp. 175-216.

3

Canonical integration and analysis of periodic maps using non-standard analysis and Lie methods

Etienne Forest and Martin Berz

ABSTRACT We describe a method and a way of thinking which is ideally suited for the study of systems represented by canonical integrators. Starting with the continuous description provided by the Hamiltonian, we replace it by a succession of preferably canonical maps. The power series representation of these maps can be extracted with a computer implementation of the tools of non-standard analysis and analyzed by the same tools. For a nearly integrable system, we can define a Floquet ring in a way consistent with our needs. Using the finite time maps, the Floquet ring is defined only at the locations s_i where one perturbs or observes the phase space. At most the total number of locations is equal to the total number of steps of our integrator. We can also produce pseudo-Hamiltonians which describe the motion induced by these maps.

3.1 Introduction: The equation of motion

The work presented here is motivated by the study of single particle motion in a complex circular storage ring. If we denote by $s(0 < s < 1)$, the position around this ring, we can define a map $\mathcal{G}(s; s+1)$ which describes completely our system. The map $\mathcal{G}(s; s+1)$ propagates any function of phase space f belonging to the set $\mathcal{V}$ of analytic functions into its new form after one iteration (or one turn). It is well known that the map $\mathcal{G}$ preserves the Poisson bracket. It is a symplectic map:

$$f \in \mathcal{V} \xrightarrow{\mathcal{G}} \mathcal{G}f \in \mathcal{V}\,; \quad \mathcal{G} \in \mathrm{End}(\mathcal{V}) \tag{1.1a}$$

$$\mathbf{x} \in \mathbf{R}^{2N} \xrightarrow{f} f(\mathbf{x}) \in \mathbf{R} \tag{1.1b}$$

$$\mathbf{x} = (q_1, p_1, \ldots, q_N, p_N) \tag{1.1c}$$

where $\mathbf{q}$ are the positions and $\mathbf{p}$ the momenta. We choose canonical variables which obey the famous Poisson bracket condition

$$[q_i, p_j] = \delta_{ij}\,. \tag{1.1d}$$

As we just said, the Poisson bracket is preserved by $\mathcal{G}$:

$$(f,g) \in \mathcal{V}^2 \xrightarrow{[\,,\,]} [f,g] \in \mathcal{V}, \tag{1.2a}$$

$$\mathcal{G}[f,g] = [\mathcal{G}f,\mathcal{G}g]. \tag{1.2b}$$

Furthermore, we assume that our starting point is a Hamiltonian $H(t)$ describing the motion between t and $t+dt$:

$$\frac{d}{dt}f(\mathbf{x})|_{\mathbf{x}=\mathbf{x}(t)} = [-H(t), f(\mathbf{x})]|_{\mathbf{x}=\mathbf{x}(t)}. \tag{1.3}$$

We can derive a differential equation for the map $\mathcal{G}(s;s+t)$ in terms of the Lie operator $:-H(t):$ associated with $H(t)$[1],[2]:

$$\frac{d}{dt}\mathcal{G}(s;s+t) = \mathcal{G}(s;s+t):-H(t):, \tag{1.4a}$$

$$\mathcal{G}(s;s) = \mathcal{E}\,; \quad \mathcal{E} = \text{identity}, \tag{1.4b}$$

$$:f:g = [f,g]\,; \quad :f:\in \text{End}(\mathcal{V}). \tag{1.4c}$$

The fundamental problem of particle optics in a periodic system is the understanding of the effect of repeated iteration of $\mathcal{G}$. In other words, we would like to understand the map $\mathcal{G}(s;s+n) = \mathcal{G}(s;s+1)^n$ as the integer n gets larger and larger.

The goal of this paper is to present a new powerful set of techniques which allow us to perform some standard normalization transformations on the map $\mathcal{G}(s;s+1)$ without compromising on the actual complexity of the original Hamiltonian $H(t)$. To put this in perspective, we will first describe the usual way these studies are done by accelerator physicists.

3.2 Conventional approach for the study of $\mathcal{G}(s;s+1)$

3.2.1 Ray tracing

The simplest way to include all the information contain in $H(t)$ is to integrate Eq.(1.3) for the $2N$ projection functions Π_i defined as follows:

$$\mathbf{R}^{2N} \xrightarrow{\Pi_i} \mathbf{R}; \quad \Pi_i \in \mathcal{V}, \tag{2.1a}$$

$$\Pi_i(\mathbf{x}) = x_i\,. \tag{2.1b}$$

Of course, only a finite set of initial conditions can be traced and therefore it is hard to extract a lot of information from ray tracing. In this brute force approach, the simulation can be done with a Hamiltonian $H(t)$ which can have all the pieces simulating field errors, position errors or any other messy effects. Traditionally and accidentally, accelerator physicists have used low order explicit symplectic (or canonical) integrators to simulate their complex machines.

What if analytical results are needed? How do they proceed? The answer is simple: they go back to the Hamiltonian $H(t)$. We briefly describe this approach is the next paragraph.

3.2.2 NORMALIZATION OF $H(t)$

During a normal form process, one attempts to transform the Hamiltonian $H(t)$ into a "simpler" Hamiltonian $K(t)$ by a periodic canonical transformation $\mathcal{A}(t)$. This process requires the knowledge of $\mathcal{A}(t)$ for all t's. For example, let us suppose that we have the following relations:

$$\mathbf{z} = \mathcal{A}(t)^{-1}\mathbf{x} = \exp(:-w(\mathbf{x};t):)\,\mathbf{x}, \tag{2.2a}$$

$$\mathbf{z} = \text{new variables}\,. \tag{2.2b}$$

Then the new Hamiltonian is given by a relation first derived by Cary (1978)[3]:

$$K(t) = \mathcal{A}(t)\left(H(t) + \operatorname{iexp}(:-w(t):)\,\frac{\partial w}{\partial t}\right), \tag{2.3a}$$

$$\operatorname{iexp}(\lambda) = \frac{\exp(\lambda)-1}{\lambda}\,. \tag{2.3b}$$

Corresponding to $K(t)$, a map $\mathcal{N}(t)$ can be derived and it obeys Eq.(1.4). In fact, using the definition of $\mathcal{A}$ and remembering that our maps transform functions of phase space, we get:

$$\mathcal{N}(s;s+1) = \mathcal{A}(s)\,\mathcal{G}(s;s+1)\,\mathcal{A}(s)^{-1}\,. \tag{2.4}$$

If the system under study is of sufficient complexity, the direct computation of $K(t)$ and $\mathcal{A}(t)$ is not feasible without approximating $H(t)$. Hence a Birkhoff (or Deprit) type normalization of the map described by $H(t)$ is rarely done in a way consistent with the results of the integrator used in tracing rays in phase space.

3.2.3 CONCLUSION ON OLD METHODS

Perturbation theory on the Hamiltonian is not feasible on complex systems. In addition, since we are interested in the behavior of $\mathcal{G}(s;s+1)$ at a finite number of locations s, the complete solution required by Eq.(2.7) contains a lot of useless information even when obtained.

What can be done to maintain the generality of the integrator and at the same time perform normalization algorithms on the very model used by the integrator without further approximations? This we will answer in the next sections.

3.3 A new approach for the study of $\mathcal{G}(s; s+1)$

What we have done and will now describe is to adapt old methods and develop new ones in order to match exactly the hard and easy aspects of the systems we intend to study. The tools that will be invoked in this section are not necessarily more complicated than the old tools, but they are more suited to the study of complex periodic systems.

We list the important concepts in our approach in the chronological order in which they appear in a calculation.

i) An explicit integrator must be written to describe the system. Whenever possible, we prefer to use an explicit canonical integrator. Integrators of this type have been derived for Lie groups for up to fourth order in the time step. Without further approximations, rays can now be traced.

ii) The following is true for any symplectic map and in particular for the one turn map $\mathcal{G}$, given any function $f \in \mathcal{V}$:

$$(\mathcal{G}(s; s+1)f)(\mathbf{x}) = f(\mathbf{z}) \tag{3.1a}$$

$$\begin{aligned} \mathbf{z}(s) &= (z_1(s), \ldots, z_{2N}(s)) \\ &= ((\mathcal{G}(s; s+1)\Pi_1)(\mathbf{x}), \ldots, (\mathcal{G}(s; s+1)\Pi_{2N})(\mathbf{x})). \end{aligned} \tag{3.1b}$$

we need only to compute the $2N$ functions $z_i(\mathbf{x})$ to know the action of the full map $\mathcal{G}$ on any function belonging to $\mathcal{V}$. Property (3.1) can be viewed as a consequence of Hamilton's equations Eq.(1.3) or a consequence of the differential character of Lie operators [they obey Leibnitz rule: $:f:(gh) = (:f:g)h + g(:f:h)$].

Furthermore, since our goal is to perform a perturbative calculation ordered by the degrees in the power series expansion of H around a periodic orbit $\mathbf{z}^0(s)$ ($\mathbf{z}^0(s+1) = \mathbf{z}^0(s)$), we will need to obtain the power series of the $2N$ functions z_i to a predetermined order N_o around the periodic orbit.

This particular calculation can be done with the new software implementation of the powerful tools of differential algebra which itself is an application of non-standard analysis. Essentially, we create a new field $\mathcal{R}$ which is an extension of $\mathbf{R}$. An analytic function over the field $\mathbf{R}$ is known completely if it is known for one single point of the field $\mathcal{R}$ (super-Cauchy theorem!). Hence our integrator is instructed to compute the projection functions Π_i by following the periodic orbit $\mathbf{z}^0(s)$ in that super-field. In practice (and in FORTRAN![5]), one has augmented the integrator introduced in item *i*) with a map extraction algorithm which will provide a power series expansion of the functions $z_i(s)$ around any orbit and in particular around a periodic orbit. These tools are limited in the order N_o and in the number of degrees of freedom $2N$ by the power of our computers only. The coefficients of the resulting power series are exactly computed for (and by) our integrator and are therefore very accurate as compared to those obtained by numerical differentiation. In fact the only source of inaccuracies in the computation of these coefficients is in the cumulative truncation error produced by a large number of operations.

To sum up we have cheated the gigantic size of our task twice: firstly we noticed that the action of the infinite dimensional Lie group of classical dynamics can be studied by its action on the projection functions Π_i and secondly we constructed a super-field in which these functions need to be evaluated on one element only.

In addition, once a particular ordering of the Lie operators [4] has been chosen to represent a symplectic map then the following statements are true:

Ideally two different elements of the super-field $\mathcal{R}$ representing symplectic maps will lead to different symplectic maps. In practice two sets of symplectic functions Π_i lead to the same Lie representation if their power series are identical to the highest order considered (*i.e.* N_o). Hence the functions Π_i belong to an equivalence class which is independent of the reconstruction process. For a non-canonical integrator, this is not true. For each process there exists a infinite number of non-symplectic functions leading to the same Lie operators. Hopefully the process is chosen so as to introduce an error which is not greater than the violation of the symplectic condition in the integrator.

iii) The production of the map and the analysis of the one period map are now independent procedures. The analysis and normalization of this map are best achieved by computing from the functions $z_i(s)$ the Lie generators of the original map $\mathcal{G}(s; s+1)$. This is not surprising since the process leading from $H(t)$ to $K(t)$ was perform on these Hamiltonians which are just the infinitesimal generators of $\mathcal{G}$ and $\mathcal{N}$ as indicated by Eq.(1.4). This process can also be handled by the software tools in the differential algebra package.

iv) Finally, if $\mathcal{G}(s; s+1)$ represents a quasi-integrable system, which we want to perturb at a few selected locations by a very nonlinear force, we can set up a Hamiltonian-free context for that study. More precisely, we can derive a infinite number of pseudo-Hamiltonians which describe the quasi-integrable system between the locations of the additional nonlinear force.

In sections 4, 5, 6 and 7 we will describe in a moderate amount of detail the four items on this list which allow us to integrate complex systems and properly analyse them.

3.4 Canonical integration in the symplectic group

3.4.1 EXPLICIT INTEGRATION IN A LIE GROUP

We will now describe a very ideal situation which occurs most of the time in the study of medium to large periodic systems. In these systems, for reasons outside the range of this paper, the Hamiltonian $H(t)$ and its Lie operators can be broken in two terms which are exactly solvable in terms of simple functions:

$$H(t) = H_1(t) + H_2(t)\,. \tag{4.1b}$$

$$:-H(t): \;=\; :-H_1(t):+:-H_2(t):\,. \tag{4.1b}$$

In other words, we have an explicit representation for the action of $\mathcal{M}_i$ on the coordinate projection functions Π_i. Of course, $\mathcal{M}_i(t)$ is a solution of Eq.(1.4a):

$$\frac{d}{dt}\mathcal{M}_i(s;s+t) = \mathcal{M}_i(s;s+t)\;:-H_i(t):\,. \tag{4.2}$$

How to combine the two solvable maps $\mathcal{M}_i$ over a time step Δt and approximate the full map $\mathcal{G}(s;s+\Delta t)$ is the fundamental question behind "two-map" explicit symplectic integrators. Needless to say that the problem stated here can be generalized to any Lie group by simply stating that the three operators $:-H:$, $:-H_1:$, and $:-H_2:$ belong to the Lie algebra of the group under study.

First, it is often possible to get rid of the t-dependence by a temporary extension of phase space:

$$:-h: \;=\; :-H_1(t)-p_t:+:-H_2(t)-p_t:+:p_t:, \tag{4.3a}$$

$$= \;:-h_1:+:-h_2:+:p_t:\,. \tag{4.3b}$$

The new "time-like" variable τ is related to t by the relation:

$$\frac{dt}{d\tau} = \frac{\partial h}{\partial p_t} = 1; \quad \text{hence} \Rightarrow \; \tau = t\,. \tag{4.4}$$

Using the τ-indepence of $:-h:$, we can immediately write an exact formal solution for the maps $\mathcal{G}(s;s+\Delta t)$ and $\mathcal{M}_i(s;s+\Delta t)$:

$$\mathcal{G}(s;s+\Delta t) := \exp(-\Delta t:h:)\,, \tag{4.5a}$$

$$\mathcal{M}_i(s;s+\Delta t) := \exp(-\Delta t:h_i:)\,. \tag{4.5b}$$

We use the symbol := to indicate that the maps are not really equal since the original one acts on functions of only $2N$ variables while the new map acts on the extended phase space of $2N+2$ variables. Making use of the

Campbell-Baker-Hausdorff formula and of Eqs.(4.5a) and (4.5b), we can derive four approximations of $\mathcal{G}(s;s+\Delta t)$:

$$\begin{aligned}
\exp(-\Delta t : h :) &= \exp(-\Delta t : h_1 :)\exp(-\Delta t : h_2 :)\exp(\Delta t : p_t :) \\
&\quad + \cdots \text{order}(\Delta t^2)\,, \\
\mathcal{G}(s;s+\Delta t) &= \mathcal{M}_1(s;s+\Delta t)\mathcal{M}_2(s+\Delta t;s+2\Delta t) \\
&\quad + \cdots \text{order}(\Delta t^2)\,; \qquad (4.6a) \\
\exp(-\Delta t : h :) &= \exp(-\tfrac{1}{2}\Delta t : h_1 :)\exp(\tfrac{1}{2}\Delta t : p_t :)\exp(-\Delta t : h_2 :) \\
&\quad \times \exp(\tfrac{1}{2}\Delta t : p_t :)\exp(-\tfrac{1}{2}\Delta t : h_1 :) \\
&\quad + \cdots \text{order}(\Delta t^3), \qquad (4.6b) \\
\mathcal{G}(s;s+\Delta t) &= \mathcal{M}_1(s;s+\tfrac{1}{2}\Delta t)\mathcal{M}_2(s;s+\Delta t)\mathcal{M}_1(s+\tfrac{1}{2}\Delta t;s+\Delta t) \\
&\quad + \cdots \text{order}(\Delta t^3). \qquad (4.6c)
\end{aligned}$$

A fourth order integrator can also be derived, however due to its lengthy expression, we give only the time independent result [6]:

$$\begin{aligned}
\mathcal{G}(\Delta t) &= \mathcal{M}_1(s_1)\mathcal{M}_2(d_1)\mathcal{M}_1(s_2)\mathcal{M}_2(d_2)\mathcal{M}_1(s_2)\mathcal{M}_2(d_1)\mathcal{M}_1(s_1) \\
&\quad + \cdots \text{order}(\Delta t^5)\,; \qquad (4.7a)
\end{aligned}$$

$$\begin{aligned}
s_1 &= \frac{1}{2(2-\beta)}\Delta t\,, & (4.7b) \\
s_2 &= \frac{(1-\beta)}{2(2-\beta)}\Delta t\,, & (4.7c) \\
d_1 &= \frac{1}{2-\beta}\Delta t\,, & (4.7d) \\
d_2 &= \frac{-\beta}{2-\beta}\Delta t\,, & (4.7e) \\
\beta &= 2^{1/3}\,. & (4.7f)
\end{aligned}$$

3.4.2 Implicit integration in a Lie group

For the symplectic group it is possible to write an implicit symplectic integration scheme using a characteristic function. This works all the time even when the Hamiltonian $H(t)$ precludes the existence of an explicit solution. For example, we can reproduce the effect of $\mathcal{G}(s;s+\Delta t)$ on the projection functions Π_i to first order in Δt by tracking using the characteristic function:

$$F_1 = \mathbf{q}(s)\cdot\mathbf{p}(s+\Delta t) - \Delta t\ H(\mathbf{q}(s),\mathbf{p}(s+\Delta t);s)\,. \qquad (4.8)$$

In general F_k is a solution of the Hamilton-Jacobi equation [7]:

$$\frac{\partial F_k}{\partial t} + H(\mathbf{q},\frac{\partial F_k}{\partial \mathbf{q}};t) = 0 + \cdots \text{order}(H^{k+1})\,. \qquad (4.9)$$

The map extraction techniques described in the next section are still applicable to a characteristic function integrator. First, one computes the central trajectory by solving the implicit set of equations produced by F_k and then F_k is expanded around this trajectory and the resulting power series is partially inverted using the the differential algebra tools of section 5.

3.4.3 Conclusion

We have seen how one produces simple symplectic integrators of the explicit and implicit kind. Furthermore, if only a map is needed for analysis or tracking, it can be obtained from a non-canonical integrator. The resulting power series can be "symplectified" by a procedure which extracts a characteristic function or a Lie operator representation of $\mathcal{G}$. These computations are all done by the tools of the next two sections.

3.5 Non-standard analysis and its application to map extraction

In this section we will discuss a very powerful tool for the advertised study of the production and analysis of maps in their power series representation. For our practical purposes, it will allow a very straightforward computation of derivatives and thus Taylor series expansions for very complicated functions of arbitrarily many variables to arbitrary order on a computer. It is based on a rigorous treatment of infinitely small quantities, differentials, and will allow the computation of derivatives from the difference quotient using differential differences.

Contrary to the other existing methods for rigorous treatment of such differentials (see for instance references [8], [9]), our method is fully constructive and can be implemented on a computer.

We start by defining a special collection of subsets of the real numbers, the collection of almost-finite sets $\mathcal{F}$. We call a set almost-finite, if for any real number r there are only finitely many elements of the set which are smaller than r. From this definition it is clear that all finite subsets of the real numbers are in $\mathcal{F}$. Furthermore, it is easy to convince oneself that each subset of the positive numbers is in $\mathcal{F}$. However, the set of all integers is not in $\mathcal{F}$, since there are infinitely many integers smaller than zero.

It is easy to show that a subset of any set in $\mathcal{F}$ is also in $\mathcal{F}$, and that with two sets their union and their intersection are in $\mathcal{F}$. Furthermore, with two sets $M, N \in \mathcal{F}$ the set $M + N = \{x + y \mid x \in M, y \in N\}$ is in $\mathcal{F}$, and there are only finitely many ways to write an element of $M + N$ as a sum of two elements of M and N, respectively. Finally we note that each set in $\mathcal{F}$ has a smallest element.

We are now ready to define a new, very large set, which will turn out to be a generalization of the real numbers and also contain "infinitely small" numbers. We define $\mathcal{R}$ to be the set of functions on the real numbers that are zero everywhere except on a set which is almost-finite, *i.e.* the set of non-zeroes of such functions belongs to $\mathcal{F}$.

On $\mathcal{R}$ we define an addition by just adding the functions. The resulting sum function is again in $\mathcal{R}$ because the set of non-zeroes of the sum function is contained in the union of the non-zeroes of the functions to be summed, and is hence in $\mathcal{F}$ according to the above reasoning. Hence the set $\mathcal{R}$ is closed under addition.

We also define a multiplication in the following way. For two functions $f, g \in \mathcal{R}$ let N_f, N_g denote the set of non-zeroes. We define the product function $f \cdot g$ in the following way. In case $x \notin N_f + N_g$, we say $(f \cdot g)(x) = 0$. In case $x \in N_f + N_g$, we define $(f \cdot g)(x)$ in the following way:

$$(f \cdot g)(x) = \sum_{x_f + x_g = x} f(x_f) \cdot g(x_g) \tag{5.1}$$

There are only finitely many terms in the sum because as stated above, there are only finitely many ways to write an element of $N_f + N_g$ as a sum of an element of N_f and an element of N_g. Since $N_f + N_g \in \mathcal{F}$, we can infer that the product function $f \cdot g$ is again in $\mathcal{R}$.

It can be shown that with this definition of an addition and a multiplication, the set $\mathcal{R}$ becomes a field. Most of the properties of fields can be shown quite easily to be fulfilled; the only exception is probably the existence of multiplicative inverses. For details consult reference [10]. We note that the neutral element of addition is just the function identical to zero, and the neutral element of multiplication is the function which vanishes everywhere except at $x = 0$ where it has the value 1.

On this new field we introduce an ordering relation in the following way. As customary for functions, we say $f = g$, if the two functions agree for all values of x. In case $f \neq g$, we look at the smallest x where $f(x) \neq g(x)$ and denote it by $x_{\neq}$. It exists since the set of x values where f and g disagree is contained in $N_f \cup N_g$ and is hence in $\mathcal{F}$ and thus has a smallest value. We say now that $f > g$ if $f(x_{\neq}) > g(x_{\neq})$ and $f < g$ if $f(x_{\neq}) < g(x_{\neq})$.

It follows that for arbitrary f, g we have exactly one of the conditions $f = g$, $f < g$ or $f > g$. Hence the set $\mathcal{R}$ is well ordered. Furthermore, one obtains $f < g \Rightarrow f + h < g + h$, and $f < g, h > 0 \Rightarrow f \cdot h < g \cdot h$.

After we have discussed the basic properties of this new set $\mathcal{R}$, we owe the reader a justification for the introduction and usefulness of this new structure. First we note that the set of real numbers can be embedded into $\mathcal{R}$ by identifying the real number r with the function which is zero everywhere except at zero where it takes the value r. Denoting this embedding map from the real numbers into $\mathcal{R}$ by π, we can easily verify that

$\pi(x+y) = \pi(x)+\pi(y)$, $\pi(x \cdot y) = \pi(x) \cdot \pi(y)$ and that $x < y \Rightarrow \pi(x) < \pi(y)$. So π preserves the field operations $+$ and $\cdot$ and the ordering $<$.

Hence our new structure contains all the real numbers. However, it contains much more than the real numbers, especially infinitely small and infinitely large quantities as we will show now. First we denote with d the element of $\mathcal{R}$ which vanishes everywhere except at 1 where it has the value 1. By using the definition of the multiplication, we can convince ourselves that its multiplicative inverse d^{-1} is the function that vanishes everywhere except at -1 where it has the value 1. Using our ordering relations, we can conclude that for every positive real number r we have the properties $0 < d < r$, and $d^{-1} > r$. Hence d lies between zero and any arbitrary positive real number and is thus "infinitely small". On the other hand, d^{-1} is "infinitely large". Hence the number d is what physicists like to think of as a differential.

As we will see, the field $\mathcal{R}$ allows us indeed a fully rigorous treatment of infinitely small and infinitely large quantities and naturally allows the introduction of delta-functions. But most importantly for our purposes, it is helpful for the computation of derivatives and hence Taylor series expansions. We shall illustrate this with a little example. Consider the function $f(x) = x^2 + 2$. Then obviously, its derivative at $x = 2$ is 4. Having infinitely small quantities at our disposal, it is natural to try to obtain the derivative as the difference quotient

$$\frac{f(2+d) - f(x)}{d} \tag{5.2}$$

Using the arithmetic on $\mathcal{R}$ and in particular the field properties, we obtain

$$\frac{f(2+d) - f(x)}{d} = \frac{((2+d)^2 + 2) - 6}{d} = \frac{6 + 4d + d^2 - 6}{d} = 4 + d \tag{5.3}$$

Thus we obtained the exact value of the derivative, up to the infinitely small quantity d. If all we are interested in is the real derivative, we are done because we simply have to extract the "real part" from the result. So in a way we have pushed numerical differentiation techniques to the extreme; usually the error in representing the derivative by a difference quotient becomes smaller and smaller the smaller the value of Δx. In our case, Δx became infinitely small, and so did the error.

Using the extended real numbers, one can construct extended complex numbers $\mathcal{C}$ in the usual way by introducing ordered pairs and operations on them. As it turns out, very many functions on the complex numbers can be generalized to the extended field $\mathcal{R}$ in a natural manner. In particular, this holds for all analytic functions. We shall sketch this and refer to reference [10] for details. Suppose, f is analytic in a neighborhood of z. Then f can be expressed as a power series around z with a nonzero radius of convergence. It can be shown that this power series converges even for every element in $\mathcal{C}$ to an element of $\mathcal{C}$ once the argument lies within the radius of convergence.

One of the most striking properties of such analytic functions on the extended complex numbers $\mathcal{C}$ shall be illustrated now. As we know from basic Cauchy theory, every analytic function in a simply connected region in the plane is uniquely determined by its values along a closed curve completely inside the region, because then the Cauchy formula allows its computation everywhere as an integral. Hence, because of analyticity, the function needs to be given only on a certain "one-dimensional subset" of the region. For extensions of analytic functions into $\mathcal{C}$ it turns out that they are uniquely given by their value at one suitable point, namely $z_0 + d$. This is true because of the existence of the power series:

$$f(z_0 + d) = f(z_0) + d \cdot f'(z_0) + d^2 \cdot \frac{f''(z_0)}{2} + \cdots \tag{5.4}$$

Evaluating $f(z_0+d)$ in $\mathcal{C}$ and noting that all the derivatives are complex, we obtain that $f(z_0 + d)$ (as an element of $\mathcal{C}$) vanishes everywhere except for positive integer powers of d and the values at these points are the derivatives (times the factorials).

So for practical purposes, it suffices to compute an analytic function at only one point to obtain complete information about it and especially all its derivatives. We also note that this procedure can be extended to power series of several variables, which is of importance for our computation of map expansions.

In addition to the elementary operations addition and multiplication we introduce a "left shift" operation ∂ that decreases all the non-zeroes x_i by one and multiplies the values at the non-zeroes f_i with the nonzero. It turns out that this operator is a "derivative" in that $\partial(f \cdot g) = (\partial f) \cdot g + f \cdot (\partial g)$. With this derivative, we obtain a differential algebra in the sense of reference [11]. Furthermore, if the f is a power series evaluated at the point $z_0 + d$, the operation ∂ transforms f into the derivative of the power series evaluated at $z_0 + d$.

Using this "derivative", we can compute Poisson brackets and hence can do all the manipulation of functions of phase space required in our Lie algebraic treatment.

Contrary to the other methods to introduce non-standard analysis ([8], [9]), here it is possible to implement the arithmetic on the computer. In order to do that every element is characterized by the values of non-zeroes x_i and the corresponding f_i up to a certain depth of x_i. Then the operations addition, multiplication and "derivative" can be implemented following the definitions of the operations.

If the objective is to compute only power series of functions of one variable, this is even quite straightforward. The situation becomes more difficult, however, in the case of power series of many variables in which we are interested here. In this case an efficient implementation of the multiplication requires a quite elaborate storing and retrieval technique. For details we refer to reference [5].

3.6 Normal form procedures on a power series map

3.6.1 A FIRST ORDER CALCULATION ON THE LIE REPRESENTATION

As mentioned in Eq.(2.4), the normal form procedures transform the map $\mathcal{G}(s;s+1)$ into a simpler map $\mathcal{N}(s;s+1)$ (dropping s from now on):

$$\mathcal{N} = \mathcal{A}\,\mathcal{G}\,\mathcal{A}^{-1}, \tag{6.1a}$$

$$\mathcal{A} = \exp(:F_\Omega:)\cdots\exp(:F_\omega)\cdots\exp(:F_0:). \tag{6.1b}$$

This factorization of $\mathcal{A}$ was first used by Dragt and Finn in a modification of the Deprit normal form algorithm [12] (see also reference [3] for a good review of these various methods). We will also assume that $\mathcal{N}$ is a product of a linear map $\mathcal{R}$ and a nonlinear correction $\mathcal{N}_\Omega$:

$$\mathcal{N} = \mathcal{R}\,\mathcal{N}_\Omega, \tag{6.2a}$$

$$\mathcal{N}_\Omega = \exp(:T_0:)\cdots\exp(:T_\Omega:). \tag{6.2b}$$

In a perturbative process carried order by order in the canonical variable $\mathbf{x}$, the Lie polynomials T_ω and F_ω are of degree $\omega+2$. The highest order Ω will be just N_0+1 where N_0 is the highest degree of the power series of the projection functions Π_i. The map $\mathcal{R}$ will be a generalized rotation and has the following Lie operator:

$$\mathcal{R} = \exp(:f_2:), \tag{6.3a}$$

$$f_2 = \sum_{k=1}^{N} -\frac{\mu_k}{2}\left(q_k^2 + (\epsilon_k - \bar{\epsilon}_k)p_k^2\right) = \sum_{k=1}^{N} f_2^k, \tag{6.3b}$$

$$\begin{cases} \epsilon_k = 1,\ \bar{\epsilon}_k = 0, & \text{for stable motion in } k^{th} \text{ plane,} \\ \epsilon_k = 0,\ \bar{\epsilon}_k = 1, & \text{for unstable motion in } k^{th} \text{ plane.} \end{cases} \tag{6.3c}$$

Essentially we assume that the linear part of $\mathcal{G}$ can be normalized and the result is hyperbolic or elliptic motion for all the planes (q_i, p_i). To give a flavour of the calculations, we now perform a first order determination of $\mathcal{A}$ on a map $\mathcal{M}$ of the form:

$$\mathcal{M} = \mathcal{R} \exp(: \alpha f :), \tag{6.4a}$$

$$\mathcal{M} \in \mathrm{End}(\mathcal{V})\;; \qquad \mathcal{R} \in \mathrm{End}(\mathcal{V}). \tag{6.4b}$$

Consider a canonical transformation $\mathcal{A}$ whose purpose is to modify $\mathcal{M}$ into a new factorized representation $\mathcal{N}$ defined to first order in α. Using a Lie representation for $\mathcal{A}$, we get for $\mathcal{N}$

$$\begin{aligned}
\mathcal{N} &= \mathcal{A}\mathcal{M}\mathcal{A}^{-1} \\
&= \exp(: \alpha F :) \mathcal{R} \exp(: \alpha f :) \exp(: -\alpha F :) \\
&= \mathcal{R} \exp(: \alpha \mathcal{R}^{-1} F :) \exp(: \alpha f :) \exp(: -\alpha F :) \\
&= \mathcal{R} \exp\big(: \alpha\{-(\mathcal{E} - \mathcal{R}^{-1})F + f\} + O(\alpha^2) : \big) \qquad (6.5)\\
\mathcal{E} &= \text{identity map} \in \mathrm{End}(\mathcal{V})
\end{aligned}$$

Denoting by $\mathcal{T}$ the operator $\mathcal{E} - \mathcal{R}^{-1}$, it is clear from Eq. (6.5) that one must study the range and the kernel of $\mathcal{T}$ in order to specify what possible linear terms in α can remain in Eq. (6.5). Suppose f is decomposed as follows:

$$f = f_r + f_o, \qquad f_r \perp \mathrm{Ker}\,\mathcal{T} \tag{6.6}$$

Then, we can select $\mathcal{A}$ or F such that $\mathcal{N}$ becomes $\exp(: \alpha f_o + O(\alpha^2) :)$. The function F is just given by:

$$F = \mathcal{T}^{-1} f_r\,. \tag{6.7}$$

From this short discussion one sees the central importance of the map $\mathcal{R}$. The eigenvectors of $\mathcal{R}$ of unit eigenvalue will constitute the kernel Ker $\mathcal{T}$ so critical to the inversion of $\mathcal{T}$. In the case of Eq.(6.3) the operators $: f_2^k :$ form a semi-simple algebra, we purposely neglect the case $\epsilon_k = \bar{\epsilon}_k = 0$. Its inclusion would complicate the discussion, since it is not true anymore that the vector space of polynomial functions is a direct sum of the range Im $\mathcal{T}$ and the kernel Ker $\mathcal{T}$ if the Lie Algebra of the $: f_2^k :$ has a nilpotent component. We now find the range and the kernel of $\mathcal{T}$ by constructing a suitable eigenbasis for the study of $\mathcal{R}$ in the case described by Eq.(6.3).

The evaluation of $\mathcal{T}^{-1} f_r$ requires a decomposition of f_r in eigenvectors of $: f_2 :$. These eigenvectors are easy to obtain, the answer is given by:

$$: f_2^k : h_k^{\pm} = \mp (i\epsilon_k + \bar{\epsilon}_k)\mu_k\, h_k^{\pm} = \mp \lambda_k h_k^{\pm} \tag{6.8a}$$

$$h_k^{\pm} = q_k \pm (i\epsilon_k + \bar{\epsilon}_k) p_k \tag{6.8b}$$

$$: f_2^k : = -\frac{\mu_k}{2}\, h_k^+ h_k^-\,. \tag{6.8c}$$

Using this new basis, we can easily find the kernel Ker $\mathcal{T}$. Let us define a new vector as follows:

$$|\mathbf{m}, \mathbf{n}\rangle = (h_1^+)^{m_1}(h_1^-)^{n_1} \ldots (h_N^+)^{m_N}(h_N^-)^{n_N}\,. \tag{6.9}$$

Using the differential property of the operator $: f_2 :$, we can compute the eigenvalue of $|\mathbf{m},\mathbf{n}\rangle$

$$: f_2 : |\mathbf{m},\mathbf{n}\rangle = (\mathbf{n}-\mathbf{m})\cdot\boldsymbol{\lambda}\,|\mathbf{m},\mathbf{n}\rangle\,. \tag{6.10}$$

Assuming that the λ_k's are irrational and prime amongst themselves, we conclude that

$$|\mathbf{m},\mathbf{n}\rangle \in \operatorname{Ker}\mathcal{T} \Longrightarrow \mathbf{n}-\mathbf{m} = 0\,. \tag{6.11}$$

Providing that one can easily change basis to the $|\mathbf{m},\mathbf{n}\rangle$, the computation of $\mathcal{T}^{-1}f_r$ is trivial:

$$f_r = \sum_{\mathbf{m},\mathbf{n}} A_{\mathbf{m},\mathbf{n}}\,|\mathbf{m},\mathbf{n}\rangle \tag{6.12a}$$

$$\mathcal{T}^{-1}f_r = \sum_{\mathbf{m},\mathbf{n}} \frac{A_{\mathbf{m},\mathbf{n}}}{1-\exp((\mathbf{m}-\mathbf{n})\cdot\boldsymbol{\lambda})}|\mathbf{m},\mathbf{n}\rangle\,. \tag{6.12b}$$

3.6.2 Conclusion: to higher order with the differential algebra tools

In reality, we have learned that one starts the normal form process on the representation of the map $\mathcal{G}(s; s+1)$ on the projection functions Π_i as given by their power series expansions obtained around the periodic orbit. Technically we do not have a Lie representation of the map.

However, it should be plausible to the reader that we can perform a normalization procedure on the map to an arbitrary order if the following statements are true (of course they are!):

1) The differential algebra software can take derivatives and integrals of polynomial functions. Hence it can extract all the Lie operators $\exp(: f_k :)$ from a map whose power series representation of the projection functions Π_i is known to the $(k-1)^{th}$ degree. Through differential algebra we can also multiply the coefficient of the monomials of a polynomial by an arbitrary function of the exponents of the monomials, allowing us to compute the effect of $\mathcal{T}^{-1}$ in Eq.(6.12).

2) The differential algebra software can compose functions represented by power series and therefore can perform the change of basis necessary for the production of $\mathcal{N}$. In the Π_i representation, it can compose maps to an arbitrary order avoiding the use of the Campbell-Baker-Hausdorff or the Zassenhaus formulae.

Going into the details of these operations would too lengthy and the work can be found elsewhere [13]. These calculations are mathematically equivalent to a normalization process on the Hamiltonian $H(t)$. However by separating the map extraction process from the normalization algorithm we can perform the perturbative calculations exactly for the symplectic integrator used.

3.7 The Floquet representation and its Hamiltonian-free description

3.7.1 The old way: normalizing the Hamiltonian

We assume that the Hamiltonian $H(t)$ consists of an quasi-integrable part $H_0(t)$ and a residual part $V(t)$. In the standard approach, one attempts to normalized the Hamiltonian H_0 using Eq.(2.3):

$$\begin{aligned} K_0(t) &= \mathcal{A}(t)\left(H_0(t) + \mathrm{i}\exp(:-w(t):)\,\frac{\partial w}{\partial t}\right), && (7.1a)\\ K_0(t) &= K_0(\mathbf{J},t), && (7.1b)\\ \mathbf{J} &= \tfrac{1}{2}(h_1^+h_1^-,\ldots,h_N^+h_N^-), && (7.1c)\\ \mathcal{A}(t+1) &= \mathcal{A}(t). && (7.1d) \end{aligned}$$

The motion produced by the new Hamiltonian $K_0(t)$ will be a simple $\mathbf{J}$-dependent rotation-expansion. Defining the action $\mathbf{\Phi}$ in the usual manner,

$$h_k^{\pm} = \sqrt{2J_k}\exp(\mp(i\epsilon_k + \bar{\epsilon}_k)\Phi_k) \qquad (7.2)$$

we then get the usual tori for the motion of the resulting map $\mathcal{N}(s;s+t)$:

$$\mathcal{N}(s;s+t) = \exp\Big(\int_s^{s+t} :-K_0(\tau):\,d\tau\Big) = \exp(:-\Gamma(\mathbf{J};s,s+t):)\,, \qquad (7.3a)$$

$$\mathbf{\Phi}(s+t) = \mathbf{\Phi}(s) + \frac{\partial}{\partial \mathbf{J}}\Gamma(s,s+t) = \mathbf{\Phi}(s) + \mathbf{\Delta\Phi}(s;s+t)\,. \qquad (7.3b)$$

In general the map $\mathcal{G}(s;s+t)$ which describes the motion of $H_0(t)$ from s to $s+t$ can be decomposed into three factors:

$$\mathcal{G}(s;s+t) = \mathcal{A}(s)^{-1}\,\mathcal{N}(s;s+t)\,\mathcal{A}(s+t)\,. \qquad (7.4)$$

As indicated by Eq.(7.4), a function is transformed into the Floquet space at location s by $\mathcal{A}(s)^{-1}$, then it is rotated into position $s+t$ in the Floquet space where it is finally extracted by $\mathcal{A}(s+t)$. The reader will notice that one regains Eq.(2.4) by letting $t=1$. If we assume that the action of $\mathcal{A}^{-1}$ on the projection functions Π_i produces analytic functions (*i.e.* expandable in power series which are asymptotic at the very least), then one can show that:

$$\mathcal{N}(s;s+1) = \mathcal{N}(s+t,s+t+1) \text{ for all t,} \qquad (7.5a)$$

and that:

$$\mathcal{N}(s;s+1) \text{ is the same for all choices of } \mathcal{A}(s) \qquad (7.5b)$$

The function $\Gamma(s,s+1)$ associated to the Lie operator of $\mathcal{N}(s;s+1)$ is a global invariant of the full system.

Finally we can write an expression for the full transformed Hamiltonian $K(t)$:

$$K(t) = K_0(t) + \mathcal{A}(t)^{-1}V(t). \tag{7.6}$$

The freedom that we have in selecting $\mathcal{A}(t)$ is normally exploited in trying to modify $V(t)$ as little as possible in equation (7.6).

3.7.2 Conclusion on Hamiltonian normalization

Three things can be said about the procedure outlined in the previous section:

1) The solution of Eq.(7.6) requires solving for w (*i.e.* $\mathcal{A}(t)$) at every location around our complex system. Usually this is impractical for a complex system as we pointed out already.

2) Extremely nonlinear perturbations such as $V(t)$ are often localized in the variable t. Therefore only the maps between these locations are necessary. Hence the steps outlined in subsection (7.1) produce unnecessary information. In fact, if canonical integrators are used to represent $H_0(t)$ as well as $V(t)$, the map is better described by tools which do not assume any differentiability in the variable t.

3) If, for whatever reason, one needs a normalized Hamiltonian which describes correctly the map from one localized perturbation to the next then there exists an infinite number of these Hamiltonians. Not surprisingly the exact Hamiltonian $K_0(t)$ is one of them; but it is one of the most complicated choices.

In the next section, we describe an approach ideal suited for studying a finite number of localized perturbations. Ultimately, as we pointed out in section 4, an integrator is best described as a finite product of maps.

3.7.3 Floquet transformation on the map

To the extent that a symplectic integrator is a discontinuous set of operations (so is any integrator!), normalization of the Hamiltonian is not very suited to the study of the system produced by the integrator. In addition, as we just mentioned in subsection (7.2), this is even more so if a localized perturbation is added.

In this last section, we will develop an approach to the study of a full periodic system based on the maps alone [14], [15]. We will therefore assume the existence of a map $\mathcal{G}(s; s+t)$ known at few locations where we intend to perturb our system. For complete generality, we need only to consider two locations s_1 and s_2. First, we assume the existence of the maps $\mathcal{A}(s_i)$:

$$\text{there exists } \mathcal{A}_i \text{ such that } \mathcal{N} = \mathcal{A}_i^{-1}\mathcal{G}_i\mathcal{A}_i\,, \tag{7.7a}$$

$$\mathcal{N} = \exp(- : \Gamma(\mathbf{J}) :), \qquad \mathcal{G}_i = \mathcal{G}(s_i, s_i + 1)\,. \tag{7.7b}$$

As we said earlier, the uniqueness of $\mathcal{N}$ is guaranteed by the good behaviour of the power series of $(\mathcal{A}\Pi_i)(\mathbf{x})$ near the origin. In the perturbation theory of section 5, this is implicitly assumed if the power series is to have any sense at all.

To proceed further, we write $\mathcal{G}_1$ in terms of $\mathcal{G}_2$ and use the integrability relation of Eq.(7.7) twice:

$$\begin{aligned} \mathcal{G}(s_1, s_1+1) &= \mathcal{G}(s_1, s_2)\,\mathcal{G}(s_2, s_2+1)\mathcal{G}(s_1, s_2)^{-1}\,, & (7.8a)\\ \mathcal{G}(s_1, s_1+1) &= \mathcal{G}(s_1, s_2)\,\mathcal{A}_2^{-1}\,\mathcal{N}\,\mathcal{A}_2\mathcal{G}(s_1, s_2)^{-1}\,, & (7.8b)\\ \mathcal{N} &= \mathcal{A}_1\mathcal{G}(s_1, s_2)\mathcal{A}_2^{-1}\,\mathcal{N}\,\mathcal{A}_2\mathcal{G}(s_1, s_2)^{-1}\mathcal{A}_1^{-1}\,. & (7.8c) \end{aligned}$$

Finally, using the Lie representation of $\mathcal{N}$:

$$\begin{aligned} \mathcal{N} &= \mathcal{B}(s_1, s_2)\,\mathcal{N}\,\mathcal{B}(s_1, s_2)^{-1}\,, & (7.9a)\\ \mathcal{N} &= \exp(- : \Gamma(\mathcal{B}(s_1, s_2)\mathbf{J}) :)\,, & (7.9b) \end{aligned}$$

where

$$\mathcal{B}(s_1, s_2) = \mathcal{A}_1\mathcal{G}(s_1, s_2)\mathcal{A}_2^{-1}\,. \tag{7.9c}$$

From Eq.(7.9b) we conclude that $\mathcal{B}$ must leave $\mathbf{J}$ invariant. Hence it must have the form:

$$\mathcal{B}(s_1, s_2) = \exp(- : \Gamma(\mathbf{J}; s_1, s_2) :)\,. \tag{7.10}$$

The function $\Gamma(\mathbf{J}; s_1, s_2)$ is identical to its Hamiltonian counterpart of Eq.(7.3).

Something must be said about the choice of $\mathcal{A}_i$: technically the only constraint is that $\mathcal{A}_i$ must be uniquely defined for each position s_i in order to insure periodicity. In addition, we often restrict ourselves to $\mathcal{A}_i$'s which are internally defined; by this we mean that our construction method is an injection from the set of possible $\mathcal{G}(s_i, s_i+1)$ (finite set for an integrator) to the infinite set of possible $\mathcal{A}_i$. This insures that the phase advance between two "matched" locations is the same for all possible injective constructions (Matched locations have the same one turn map $\mathcal{G}$). Again however, the nature of the perturbation added to the map will dictate the choice of $\mathcal{A}$.

Finally, we add two localized perturbations at positions s_1 and s_2, given by their Lie operators $: V_1 :$ and $: V_2 :$ respectively. The new maps $\mathcal{G}^{\text{total}}(s_1, s_1+1)$ and $\mathcal{N}^{\text{total}}(s_1, s_1+1)$ are just:

$$\begin{aligned} \mathcal{G}^{\text{total}}(s_1, s_1+1) &= \exp(- : V_1 :)\,\mathcal{G}(s_1, s_2)\\ &\quad \times \exp(- : V_2 :)\,\mathcal{G}(s_2, s_1) & (7.11a)\\ \mathcal{N}^{\text{total}}(s_1, s_1+1) &= \exp(- : \mathcal{A}_1 V_1 :)\\ &\quad \times \exp(- : \mathcal{B}(s_1, s_2)\mathcal{A}_2 V_2 :)\,\mathcal{N}\,. & (7.11b) \end{aligned}$$

The various factors of $\mathcal{N}^{\text{total}}(s_1, s_1+1)$ have a simple interpretation. The first two factors represent the perturbations V_1 and V_2. One notices that each perturbation is modified by two maps: firstly, as we mentioned, the Lie operator is brought into the Floquet representation by $\mathcal{A}$; secondly the operator has its phase $\mathbf{\Phi}$ advanced by the appropriate amount by the map $\mathcal{B}(s_1, s_2)$. The third factor is the unperturbed Floquet map $\mathcal{N}$. One might view the full action of $\mathcal{N}^{\text{total}}(s_1, s_1+1)$ on an arbitrary function f as follows: first f is distorted into a new function by the perturbations and then it is transformed by the original map $\mathcal{N}$.

As we said earlier, it is sometimes useful to produce a Hamiltonian which describes the map $\mathcal{N}^{\text{total}}(s_i, s_i+1)$ for $(i = 1,2)$. Of course, the exact Hamiltonian $H(t)$ does this for all possible s:

$$H(t) = H_0(t) + \sum_{i=1,2} \delta_p(s - s_i) V_i\,, \tag{7.12a}$$

$$\delta_p(s) = \delta_p(s+m), \quad m = \text{integer}\,. \tag{7.12b}$$

However, this defeats the purpose of using maps. Indeed we assumed that the map is known only between s_1 and s_2. Nevertheless it is easy to write a infinite number of Hamiltonians which reproduce the motion of $\mathcal{N}^{\text{total}}(s_i, s_i+1)$ for $(i = 1,2)$:

$$K_\kappa(t) = \kappa(\mathbf{J}, t) + \sum_{i=1,2} \delta_p(s - s_i)\mathcal{A}_i V_i\,, \tag{7.13a}$$

where κ is such that:

$$\int_{s_1}^{s_2} \kappa(\mathbf{J}, t)\, dt = \Gamma(\mathbf{J}, s_1, s_2)$$

and

$$\int_{s_1}^{s_1+1} \kappa(\mathbf{J}, t)\, dt = \Gamma(\mathbf{J}, s_1, s_1+1)\,. \tag{7.13b}$$

The Hamiltonians of Eq.(7.13) have the proper phase advance $\mathcal{B}$ between the the points s_1 and s_2 and the correct $\mathcal{A}$ at these locations. The function κ can be chosen to be very simple (or extremely complex).

3.8 Acknowledgements

We would like to thank Drs. Swapan Chattopadhyay and Alex Chao for providing the moral and financial support for our work.

3.9 References

[1] A.J. Dragt and E. Forest, *J. Math. Phys.* **24**, 2734 (1983).

[2] The derivation of this result uses a differential equation for the symplectic map $\mathcal{M}$ and the concept of the intermediate (or Dirac) representation. It is explained in detail in reference 1. One can consult

A. Messiah *Mecanique Quantique*, (Dunod 1965) p. 270 for a familiar quantum mechanical version of the derivation.

[3] J.R. Cary, *Phys. Rep.* **79**, 129 (1981).

[4] A nonlinear map expanded into a power series can be factorized into a linear and a purely nonlinear part. One can either compute a single Lie operator for the nonlinear part or factor it into a product of Lie transformations of increasing degree. See reference 1 or the original derivation in: A.J. Dragt and J.M. Finn, *J. Math. Phys.* **20**, 2649 (1979).

[5] A description of the FORTRAN implementation of differential algebra by M. Berz is to be found in SSC-152 or SSC-166 technical reports of the Superconducting Super Collider Central Design Group. Both reports have been accepted for publication in Particle Accelerators.

[6] R. Ruth, *IEEE Trans. Nucl. Sci.* NS-30, 2669 (1985). The results of this original paper are general and applicable to any Lie group despite the lack of generality in the proofs.

[7] P. Channel, LANL Technical. Note AT-6:ATN-83-9 (1983); P. Channel and J.C. Scovel "Symplectic Integration of Hamiltonian Systems", to appear in the proceedings of the Los Alamos Symplectic Integrator workshop (1988).

[8] D. Laugwitz, "Ein Weg zur Nonstandard-Analysis", *Jahresberichte der Deutschen Mathematischen Vereinigung*, (1973) p. 66–75.

[9] A. Robinson, "Non-Standard analysis", *Proceedings Royal Academy Amsterdam*, Series A, (1961) p. 432.

[10] M. Berz, "Analysis auf einer nichtarchimedischen Erweiterung der reellen Zahlen", preprint.

[11] J.F. Ritt, *Differential Algebra* (American Mathematical Society, 1950).

[12] A.J. Dragt and J.M. Finn, *J. Math. Phys.* **17**, 2215 (1976).

[13] See reference 5 (SSC-166) for a detailed description of the differential algebra implementation of the normal form process.

[14] E. Forest, SSC Report 111 (1987) and SSC Report 138 (1987), to appear in the proceedings of the 1987 Fermilab summer school in accelerator physics.

[15] Independently of our Hamiltonian-free approach, it has been developed and used by another group. See: A. Bazzani, P. Mazzanti, G. Servizi, and G. Turchetti, "Normal Forms for Hamiltonian Maps and Non Linear Effects in a LHC Model", CERN SPS/88-2 (AMS) LHC note 66 (European Organisation for Nuclear Research, 1988).

4

Concatenation of Lie Algebraic Maps

Liam M. Healy and Alex J. Dragt

ABSTRACT Time evolution in a Hamiltonian system may be represented by a transfer map, which in turn may be represented as a product of Lie transformations factored by order. Two such products in succession may be concatenated into a single product. It is possible to do this even when the Lie transformations include inhomogeneous terms. Rules are given for combining and ordering Lie transformations in the general case through sixth order. These rules are presented in an algorithmic fashion suitable for manipulation by computer.

Such techniques have applications to many Hamiltonian systems, including accelerator beam dynamics and optics. The concatenation could represent, for instance, the combined effects of two successive beamline elements in a particle accelerator. In this case, inhomogeneous terms can arise when there are placement, alignment, or powering errors.

4.1 Introduction

In a Hamiltonian system such as a particle accelerator, it is often important to know where a particle will be at a certain time given its position in phase space at some previous time. The relation that gives this information is called the transfer map. This transfer map can be represented by a sequence of Lie transformations factored by order.

It is often desirable to know to a given order the Lie product corresponding to two successive mappings; mathematically we wish to *concatenate* the two Lie products into one: this is the same as composition of the transfer maps. Treatment of this problem is the primary goal of this paper. One may visualize this, in a particle accelerator, as a means of finding the particular Lie product for the combined effect of two successive beamline elements knowing each separately. This is useful in a variety of calculations. For example, in a single pass system, one can find the combined aberrations of a complete system in terms of the aberrations of its components. Also, when tracking a particle through many turns in a circular accelerator, a great saving in computation time will be realized if many magnetic beamline elements are lumped together as one. Finally, there is an extensive array of analytical tools [8] for analyzing the full one-turn map.

We shall not concern ourselves in this paper with the generation of the original Lie product from the Hamiltonian; that is treated in another paper [12].

4.1.1 Definitions and terminology

It is shown in another paper [12] that the transfer map describing the time evolution of a Hamiltonian system may be represented as a factored product of Lie transformations. Let the transfer map $\mathcal{M}$ send any point in phase space z to another $\bar{z}$ via the relation

$$\bar{z} = \mathcal{M}\, \zeta|_{\zeta=z}, \tag{1.1}$$

where $\zeta = (q_1, p_1, \ldots, q_l, p_l)$ are the $2l$ phase space variables, and z is a $2l$-tuple of values in phase space. Suppose we have a known solution to the time evolution for the orbit through a particular point in phase space. This solution will be called the central orbit. Then, with a suitable choice of coordinates, $\mathcal{M}$ may be represented in the neighborhood of the central orbit in the form

$$\mathcal{M} = e^{:f_1:} e^{:f_2:} e^{:f_3:} \cdots . \tag{1.2}$$

Here, f_n is a polynomial homogeneous of order n in the phase space variables. The *Lie operator* $:f:$ associated with f is defined by its action on another function g on phase space,

$$:f:g \stackrel{\text{def}}{=} [f, g]. \tag{1.3}$$

Here $[,]$ denotes the Poisson bracket. The exponential of a Lie operator is defined by the familiar Taylor series for the exponential,

$$e^{:f:} = \mathcal{I} + :f: + \frac{:f:^2}{2!} + \frac{:f:^3}{3!} + \cdots . \tag{1.4}$$

The exponential of a Lie operator is called a *Lie transformation.* A Lie transformation $e^{:f_n:}$ corresponding to a homogeneous n^{th}-order polynomial is called an nth-order transformation, and the representation of $\mathcal{M}$ as (1.2) is called the *standard factorization*. As shown in [12], the Lie product (1.2) acting on phase space is equivalent to a Taylor series. It should be pointed out that not all maps can be put in the form (1.2); some will require a second second-order transformation, see [12] or [3]. In general we use (1.2) metaphorically.

A Lie algebra is a vector space with a multiplication operation that satisfies linearity,

$$[af + bg, h] = a[f, h] + b[g, h] \quad \text{and} \quad [f, ag + bh] = a[f, g] + b[f, h], \tag{1.5}$$

antisymmetry,

$$[f, g] = -[g, f], \tag{1.6}$$

and the Jacobi identity,

$$[f,[g,h]]+[g,[h,f]]+[h,[f,g]]=0. \tag{1.7}$$

There are two primary Lie algebras we use. One is the set of functions on phase space S, of which ζ may be thought a member, with the multiplication operation being a Poisson bracket. The other is the set of Lie operators S^* with the multiplication operation being the commutator. There is a ***homomorphism***, or multiplication-preserving map, between these two spaces, called 'Ad'. Because of this homomorphism, these spaces may be thought of interchangeably, and the distinction will not be made henceforth.

The product (1.2) maps S into itself in the following fashion. First, note that the order of the Poisson bracket of two homogenous polynomials is the sum of their respective orders, minus two. That is, the Poisson bracket $[f_n, g_m]$ is homogenous of order $n+m-2$. From this we may anticipate the resultant order of applying a Lie transformation of a certain order. The ith term in the series (1.4) is $:f:/i$ applied to the previous term. This means a second-order Lie transformation $e^{:f_2:}$ is a linear transformation because each application of $:f_2:$ leaves the order of its argument unchanged. By contrast, a third-order transformation $e^{:f_3:}$ is non-linear, each application of $:f_3:$ increasing the order of its argument by one; a fourth-order tranformation $e^{:f_4:}$ is also nonlinear, increasing the order of its argument by two at each application of the operator, and so on. A first-order transformation $e^{:f_1:}$, is a special case. Since $:f_1:$ decreases by one the order of its argument, the transformation $e^{:f_1:}$ applied to a polynomial results, in general, in terms of all orders down through a constant. In particular, if it acts on the phase space functions ζ, it will introduce constant terms. These first-order transformations are the result of inhomogeneities in the Hamiltonian, and have the effect of mapping the central orbit onto some other orbit.

4.1.2 The Task of Concatenation

Suppose now that we have two maps $\mathcal{M}$ and $\mathcal{N}$ defined by

$$\begin{aligned} \mathcal{M} &= e^{:f_1:}e^{:f_2:}e^{:f_3:}\ldots \\ \mathcal{N} &= e^{:g_1:}e^{:g_2:}e^{:g_3:}\ldots, \end{aligned} \tag{1.8}$$

and we wish to join these into a single map $\mathcal{P}$,

$$\mathcal{P}=\mathcal{M}\mathcal{N}. \tag{1.9}$$

We would like to know how to find the factored form of $\mathcal{P}$, *i.e.* the h such that

$$\mathcal{P} = e^{:h_1:}e^{:h_2:}e^{:h_3:}\ldots. \tag{1.10}$$

In computing this, the number of terms is important. From a practical standpoint, we must truncate each of the products $\mathcal{M}$ and $\mathcal{N}$ at a certain order N. It is reasonable to truncate the products $\mathcal{P}$ at that same value of N. The concatenation problem (1.8–1.10) can then be stated compactly: for a given N, and polynomials $f_1, \ldots, f_N$ and $g_1, \ldots, g_N$, we wish to find the polynomials $h_1, \ldots, h_N$ such that

$$e^{:h_1:}e^{:h_2:}e^{:h_3:}\ldots e^{:h_N:} \cong e^{:f_1:}e^{:f_2:}e^{:f_3:}\ldots e^{:f_N:}\, e^{:g_1:}e^{:g_2:}e^{:g_3:}\ldots e^{:g_N:}. \tag{1.11}$$

The symbol '$\cong$' means equivalent through a certain order (in this case N); it will be explained rigorously later. In combining the polynomials, however, higher-order terms will be produced, as we shall see. Furthermore, the presence of first-order terms in the product for one map will combine with any given term of the other product to produce an effect of lower order in the resultant concatenation. These considerations are addressed in the next Section.

4.2 Ideal structure of the Lie algebra

For most realistic maps, the Lie product (1.2) will be infinite. That is, there is no n for which $f_i = 0$ for $i \geq n$. It is of course not practical to deal with an infinite number of terms, so we choose an order at which to truncate the product. This entails making an approximation. This approach is valid in many real physical situations. The Hamiltonian is analytic, so all solutions are analytic in the inital conditions. Therefore, any transfer map is analytic (see [14] chapter 10). We must find a standard by which we can replace a given Poisson bracket by an arbitrary quantity (perhaps zero) if we feel it is "too small," and we must be able to do so consistently. It is crucial not only for derivation of the concatenation formula, but also for calculation of the factored Lie transformation product (1.2) from the Hamiltonian (see [12]). The procedure is divided into two parts: the homogeneous case, where $n \geq 2$ and $m \geq 2$, and the inhomogeneous, where $n = 1$ or $m = 1$. Those not interested in the mathematical underpinnings may without loss of continuity skip to the last paragraph of this Section.

4.2.1 HOMOGENEOUS PRODUCTS

The presumption of the Lie transformation product (1.2) is that the corresponding Taylor series is convergent (see [12]). Since we are truncating this product, we want the remainder term that is omitted to be so small that it can be safely dropped. In order to do this, we take the values of each of the phase space variables to be small so that sufficiently high orders may be ignored. Specifically, let each of the phase space variables carry the small factor δ, so that powers of δ count the order of these variables. Polynomials homogeneous of order n have a factor δ^n, and will be said to have *δ-rank n*.

Since δ is small we may, when taking Poisson brackets, neglect terms with δ-rank equal to or greater than some specified value N. For example, if we choose $N = 6$, then the Poisson bracket $[f_4, g_5]$ may be ignored since it is of order 7. We presently shall give this process some rigor; before doing so however, it is necessary to introduce some definitions. In the mathematical definitions that follow, S stands for an arbitrary Lie algebra and not necessarily the Poisson bracket Lie algebra of phase space functions.

We wish to divide up S by order, consistent with the Lie algebra. To do this, we need the concepts of subalgebra and ideal. A subset $S' \subseteq S$ is a *subalgebra* if[1] $[S', S'] \subseteq S'$, where $[S', S']$ is the image of the Poisson bracket restricted to S'. A subset $S' \subseteq S$ is an *ideal* if $[S', S] \subseteq S'$. That is, for $s' \in S'$, $s \in S$, then $[s', s] \in S'$. Clearly, an ideal is also a subalgebra. Ideals will play a critical role in the approximation process by way of quotient algebras.

Approximation is made by doing computations in terms of a *quotient algebra*, which can be defined for any ideal. The ideal plays the role of the remainder, the part that is ignored, and the quotient algebra is the original algebra with "elements approximately the same" identified. Specifically, let I be an ideal in the Lie algebra S. Now define the quotient algebra S/I to be the set of equivalence classes given by the equivalence relation

$$s_1 \cong s_2 \quad \text{if} \quad s_1 = s_2 + i \quad \text{for some} \quad i \in I. \tag{2.1}$$

We denote these classes by $s+I$, where $s \in S$. Since an ideal is a subalgebra, S/I is a Lie algebra with the rules

$$(s_1 + I) + (s_2 + I) = (s_1 + s_2) + I \tag{2.2}$$

$$c(s_1 + I) = cs_1 + I \quad \text{for} \quad c \in \mathbf{R} \tag{2.3}$$

$$[s_1 + I, s_2 + I] = [s_1, s_2] + I. \tag{2.4}$$

These rules are easily seen to be consistent. If i_1, $i_2 \in I$ are arbitrary, the left side of the first rule (2.2) is

$$(s_1 + i_1) + (s_2 + i_2) = (s_1 + s_2) + (i_1 + i_2) \in (s_1 + s_2) + I, \tag{2.5}$$

since I is a vector subspace of S . A similar argument holds for the second rule. The left side of the third rule (2.4) is

$$[s_1 + i_1, s_2 + i_2] = [s_1, s_2] + [s_1, i_2] + [i_1, s_2] + [i_1, i_2]. \tag{2.6}$$

---

[1]The notation $[X, Y]$ where X and Y are sets means the set

$$\{[x, y] \mid x \in X, y \in Y\}.$$

From the definition of an ideal, we see that the consistency of the third rule is upheld.

Now we address ourselves to the question of the origin of the ideals for the approximation. They come from a *percolation*, a telescoped set of subspaces whose union fill the whole space and whose Lie bracket fulfills a certain property: for each non-negative integer i, there is a subspace $S^{(i)}$ such that

$$S^{(i)} \supseteq S^{(j)} \quad \text{for} \quad i \leq j \tag{2.7}$$

$$S^{(0)} = S \tag{2.8}$$

$$[S^{(i)}, S^{(j)}] \subseteq S^{(i+j)}. \tag{2.9}$$

If S is *graded*, that is, it is the direct sum of subspaces S_i ($i = 0, 1, \ldots, \infty$) and $[S_i, S_j] \subseteq S_{i+j}$, then it is percolated by the rule

$$S^{(i)} \stackrel{\text{def}}{=} \bigoplus_{j \geq i} S_j. \tag{2.10}$$

In our application, these subspaces will be polynomials of a certain δ-rank or higher.

It is clear from the above definition that each of the members $S^{(i)}$ of a percolation is an ideal. Let $s^{(i)} \in S^{(i)}$, $s \in S = S^{(0)}$. Then $[s, s^{(i)}] \in S^{(i)}$, and also with the arguments in the reverse order, $[s^{(i)}, s] \in S^{(i)}$. Since s and $s^{(i)}$ were arbitrary within their respective sets, $S^{(i)}$ is an ideal in S .

Let us now apply this to our particular problem. Up to now in this Section, we have let S stand for an arbitrary Lie algebra. We now restrict the definition of S so that it is the space of all functions on phase space that have power series expansions and whose power-series expansion has no first-order or constant term. Grade it with subspaces of polynomials homogeneous in a particular order of the phase space variables: let S_i be the set of all homogeneous polynomials of order $i + 2$, for $i > 0$. One may easily verify that this is a grading on S under the Poisson bracket. A particular polynomial that belongs to the subspace S_i has δ-rank $i + 2$.

Given this grading $\{S_i\}$ by polynomial order, we have the corresponding percolation given by (2.10). This gives us a sequencs of ideals $S^{(i)}$, and a sequence of quotient algebras $S/S^{(i)}$. The ideal $S^{(i)}$ consists of all power series with coefficients zero for all terms of order less than $i+2$. The quotient algebra $Q^{(i)} \stackrel{\text{def}}{=} S/S^{(i+1)}$ is a rigorous way of describing the algebra S with the approximation of "neglecting terms of order $i + 3$ and greater."

The computer code MARYLIE 3.0 [2] [6] [7], designed to perform beam dynamics calculations for accelerator physics, has no first-order polynomials and performs computations through fourth order. Speaking in terms of the formal algebra introduced above, it is computing in the Lie algebra $Q^{(2)}$. In this algebra, one must say, for example, $[f_3, g_3] = h_4$ for some specific h_4, but it makes the approximation $[f_3, g_4] \cong 0$. In the algebra $Q^{(2)}$ the values of some Poisson brackets will be in S^3. Since we are working in the quotient algebra, we may choose any member of S^3. When performing the computation by hand, we will usually choose zero. When working numerically in a computer code (the tracking part of MARYLIE), the choice of some other members may be computationally advantageous.

The homomorphism Ad carries all these definitions to the adjoint algebra S^* . The subalgebras $S^{*(i)}$ are the spaces of all Lie operators that are of order $i+2$, *i.e.*, are images of S_i under Ad, and $S^{*(i)}$ are the direct sum $\oplus_{j\geq i} S_j^*$ or the image of $S^{(i)}$ under Ad. The $S^{*(i)}$ are ideals, so the $Q^{*(i)} = S^*/S^{*(i+1)}$ are quotient algebras. Thus the adjoint algebra has the same ideal and quotient structure as the underlying algebra, as we expect, and commutators of Lie operators are set to zero (or to an arbitrary value) when the Poisson bracket of the corresponding polynomials would be of too high an order.

These quotient Lie algebras give rise to quotient groups in the group of all symplectic maps on phase space. While possibly containing terms of all orders, these maps are accurate only through order $i+1$ for $Q^{(i)}$. Within the group, it will be possible to either truncate and then multiply, or multiply and then truncate, with equal validity. Specifically, let G be the group of symplectic maps on phase space and $G^{(i)}$ $(i = 0, 1, \ldots)$ be the subgroup of these maps that has a power series expansion consisting of terms only order $i+1$ and higher, plus the identity. Then $G^{(i)}$ is a normal or invariant subgroup of G, that is, $ghg^{-1} \in G^{(i)}$, $\forall g \in G, h \in G^{(i)}$. The quotient group $H^{(i)} = G/G^{(i+1)}$ is defined as the equivalence classes given by $g_1 \cong g_2$ if $g_1 g_2^{-1} = h$ where $h \in G^{(i)}$. That is, two elements are equivalent if the power series expansions of their associated symplectic maps differ only by terms of order $i+1$ and higher. That this is in fact a group may be easily verified. These groups $H^{(i)}$ are associated with the algebras $Q^{*(i)}$.

4.2.2 INHOMOGENEOUS PRODUCTS

If a first-order transformation is present in the product (1.2), the truncation by δ-rank given above will not be correct because the δ-rank will be lowered by the first-order term. Suppose we keep terms only through δ-rank 4, and discard anything higher. Then forming, for instance, $[f_1, [g_3, h_4]]$ would yield the wrong answer: we may take $[g_3, h_4]$ as zero because the δ-rank is 5, and so our overall answer would be zero. But this is not correct; even though the inner Poisson bracket is δ-rank 5, the Poisson bracket with f_1 subsequently lowers the δ-rank to an acceptable 4. Clearly, the subspaces described in the last Section are no longer subalgebras when a first-order term is present and we must reformulate their description for correct computations in this case.

The correct treatment of first-order terms becomes evident when we consider their physical origin. As is shown in [12], the first-order transformation is proportional to the first-order term in the Hamiltonian, which in turn is proportional to an error, such as an alignment or powering error in an accelerator beamline element. It is hoped that these are small; consequently, the first-order transformation will also be small. To be specific, say that a factor of the small parameter ϵ multiplies each first-order polynomial. We may now consider how this changes the analysis of the algebra and the corresponding group given in the previous Section.

The space S must be expanded, and the set $\{S_i\}$ given a new member. Let S now stand for the set of all functions on phase space that have power-series expansions; we no longer require that the first-order term be zero. We shall still ignore constant terms since they play no role in the Lie algebra. Let S_{-1} be the space of first-order polynomials. The spaces $\{S_i\}_{i=-1,0,1,\ldots}$ are still a grading. However, we cannot construct a corresponding percolation according to (2.10) because we now have a negative i. Thus the $S^{(i)}$ are no longer ideals and we cannot form the quotient algebra. There is a corresponding destruction of the normal subgroups and quotient groups of symplectic maps.

Instead of using this grading, let us search for another way to create a percolation, and thus obtain a sequence of ideals and a sequence of quotient algebras that will correctly reflect the physics of the perturbation calculation. First, let the *ϵ-rank* of a polynomial mean the lowest order in ϵ for that polynomial. Define a second index j, $j = 0, 1, \ldots$, on the S_i that is equal to the ϵ-rank. Thus S_{ij} is a subset of S that is homogeneous of order $i+2$ in the phase space coordinates, and homogeneous of order j in ϵ. A polynomial has δ-rank i and ϵ-rank j if it belongs to S_{ij}. The only combination of i and j within these ranges that is prohibited is $i=-1$, $j=0$, the smallness requirement on first-order terms discussed above. The spaces S_{ij} are a finer grading of the spaces S_i graded by δ-rank; it is easy to see that

$$[S_{ij}, S_{kl}] \subseteq S_{i+k,j+l}. \tag{2.11}$$

We now seek a percolation constructed from the S_{ij}.

A percolation that satisfies the requirements may be formed by creating a special function of the two indices i and j. Consider the set of allowed index pairs

$$\mathbf{Z}^{2*} \stackrel{\text{def}}{=} \{-1, 0, 1, \ldots\} \times \{0, 1, \ldots\} - \{(-1, 0)\}. \tag{2.12}$$

Let a *truncation criterion* ν be any function into the non-negative integers $\nu : \mathbf{Z}^{2*} \to \mathbf{Z}^+$ with the property

$$\nu(i,j) + \nu(k,l) \leq \nu(i+k, j+l). \tag{2.13}$$

Now define the sequence of subspaces $S^{(i)}$, $i \in \mathbf{Z}^+$ as the direct sum of all S_{jk} whose value of ν is at least i:

$$S^{(i)} \stackrel{\text{def}}{=} \bigoplus_{\nu(j,k) \geq i} S_{jk}. \tag{2.14}$$

This sequence is a percolation, which may be shown in the following manner. Clearly, the first and second properties of a percolation are satisfied. To show the third (2.9), let $S_{i'j'}$ be an arbitrary element of $S^{(\nu(i,j))}$ and $S_{k'l'}$ be an arbitrary element of $S^{(\nu(k,l))}$. Then

$$S_{i'j'} \in S^{(\nu(i,j))} \quad \Rightarrow \quad \nu(i',j') \geq \nu(i,j) \tag{2.15}$$
$$S_{k'l'} \in S^{(\nu(k,l))} \quad \Rightarrow \quad \nu(k',l') \geq \nu(k,l). \tag{2.16}$$

The Poisson bracket of these two polynomial spaces is

$$\begin{aligned} [S_{i'j'}, S_{k'l'}] \;&\subseteq\; S_{i'+k',j'+l'} \subseteq S^{(\nu(i'+k',j'+l'))} \\ &\subseteq\; S^{(\nu(i',j')+\nu(k',l'))} \subseteq S^{(\nu(i,j)+\nu(k,l))}, \end{aligned} \tag{2.17}$$

by the truncation criterion (2.13) and the first property of a percolation (2.7). Since $S_{i'j'}$ and $S_{k'l'}$ were arbitrary in their respective subspaces, we may conclude the third property of a percolation holds,

$$[S^{(n)}, S^{(m)}] \subseteq S^{(n+m)}, \tag{2.18}$$

for values n, m in the image of ν. By implication, therefore, the $S^{(i)}$ as defined in (2.14) are ideals. This sequence of ideals may be used to define a sequence of quotient algebras.

Note that we have determined a satisfactory set of ideals with ν undetermined except for the condition (2.13). We now must consider specific truncation criteria. Two possibilities are $\nu(i,j) = \min(i,j)$ and $\nu = \alpha i + \beta j$ where $\alpha, \beta \in \mathbb{Z}^+$. The reader may verify that these two indeed satisfy (2.13). The former case corresponds to keeping all terms except those whose δ-rank and whose ϵ-rank each exceed a certain value. The latter excludes those whose weighted sum exceeds a certain value. This form of ν satisfies a stricter condition than (2.13), in fact, $\nu(i,j) + \nu(k,l) = \nu(i+k, j+l)$, and so we have a grading $S_{\nu(i,j)}$, which may form a percolation by (2.10). This percolation is the same as the one defined by (2.14).

Normal subgroups $G^{(\nu(i,j))}$ of the group of inhomogeneous symplectic maps G may be defined by analogy to that of the homogeneous group (see Section 4.2.1). The quotient group $H^{(\nu(i,j))} \stackrel{\text{def}}{=} G/G^{(\nu(i,j))}$ is the group of transformations that will actually be used in computations.

With all the formalism aside, we must choose a particular truncation criterion with which to proceed with the actual computations. The one that makes the most intuitive sense is $\nu(i,j) = i+j$. This will be called the *total rank*. In terms of δ-rank and ϵ-rank, this criterion says that we restrict terms to $O(\delta) + O(\epsilon) \leq N$ for some value of N. Physically, this is a realistic criterion, because it means that the deviation in central orbit caused by the error is of the same order as the perturbation around the central orbit. Thus, we can expect the same accuracy in the result. For example, the misalignment of a magnetic element in an accelerator should not typically be of greater magnitude than the spread in position of particles within the beam. It is also possible to imagine the case where a different truncation criterion needs to be used; for instance, a weighted sum of the exponents as mentioned above may be appropriate. The calculations in the remainder of this paper, however, are done with the total rank criterion.

4.3 Lie algebraic tools

With the algebraic formalism established, we now turn to the problem of solving the concatentation problem (1.11). There are two important tools used in this calculation that are presented in this Section. At the moment, it is useful to introduce some notation. It will become necessary to label a polynomial by its total rank as well as its δ-rank. In this case, a superscript with the total rank in parenthesis will be placed on the polynomial, *e.g.*, $k_3^{(4)}$. Furthermore, let the function τ give the total rank of a polynomial or operator, so that, *e.g.*, $\tau(k_3^{(4)}) = 4$. In Section 4.4.2 this notation will be generalized slightly; the superscript will identify a family of terms of which at least one is of that total rank, none are lower, and some may be higher.

4.3.1 THE EXCHANGE RULE

The first tool used in the computation is the *exchange rule*. This allows us to rewrite two Lie transformations in succession such that one transformation occurs in the opposite position,

$$e^{:f:}e^{:g:} = e^{:j:}e^{:f:} \tag{3.1}$$

or

$$e^{:f:}e^{:g:} = e^{:g:}e^{:k:}, \tag{3.2}$$

where the polynomials are not necessarily homogeneous. The function j in (3.1) is given by

$$j = e^{:f:}g \tag{3.3}$$

and the function k in (3.2) is given by

$$k = e^{:-g:}f. \tag{3.4}$$

For proofs, see [4] (Theorem 3), [2], or [3] (equation 5.52).

It is also possible to bring a Lie transformation inside a function; that is,

$$e^{:f:}g(\zeta) = g(e^{:f:}\zeta), \tag{3.5}$$

for arbitrary functions f, g. This is proved in [3] and in [2]. In combination with the exchange rule, we have a powerful result because it means that an operator can be moved to the left or right of another operator by transforming that operator's argument. This is especially useful where the first operator is of δ-rank 2, in which case it produces a linear transformation of the other operator's argument.

4.3.2 COMBINING TRANSFORMATIONS AND FACTORING A SINGLE TRANSFORMATION

The other useful tool has two parts. First, we shall need to combine two (or more) transformations into a single exponent. This is accomplished using the Baker-Campbell-Hausdorff (BCH) formula. The BCH theorem says that the logarithm of the product of two exponentials is in the Lie algebra; the formula gives the explicit form. That is, if

$$e^{:f:}e^{:g:} = e^{:h:}, \tag{3.6}$$

then h is in the Lie algebra spanned by f and g. The specific formula through three brackets (see [15] or [9]) is

$$h = f + g + \tfrac{1}{2}[f,g] + \tfrac{1}{12}[f,[f,g]] + \tfrac{1}{12}[g,[g,f]] - \tfrac{1}{24}[f,[g,[f,g]]] + \cdots. \tag{3.7}$$

The second part is to rewrite a single transformation as a product of transformations factored in the proper order,

$$e^{:j_1 + j_2 + \cdots + j_n:} = e^{:k_1:}e^{:k_2:}\ldots e^{:k_m:}. \tag{3.8}$$

We do not know yet what m is, except that it will be at least n, and of course it will be no greater than N. This shall be called the *separation procedure*. It consists of two steps; we first repeatedly use the BCH formula to combine the k, represented only symbolically at this point, into a single exponent, then use a general method to find the k in terms of the j.

Consider now the general problem of *Lie function inversion*; we have a set of functions F_i of unknown polynomials k_i; we wish to solve the set k_i in terms of the known values j_i of F_i. That is,

$$j_i = F_i(k_1, \ldots, k_m) \quad \text{for} \quad i = 1, \ldots, N. \tag{3.9}$$

In our particular application, the functions F_i come from multiple application of the BCH formula; we use it to express the j_i in terms of the k_i and their Poisson brackets.

The method of inversion presented here requires only that the functions F_i satisfy two properties. The first is that each function F_i has the term k_i alone, not in a Poisson bracket; for instance $F_3(k_1, \ldots, k_m) = k_3 + [k_1, k_4] + \cdots$. The second is that the ϵ-rank of both j_1 and j_2 be at least one, *i.e.*, $\tau(j_1) \geq 2$ and $\tau(j_2) \geq 3$. The requirement on j_1 is simply the ϵ-rank requirement embodied in (2.12) and insures that the total rank is never lowered. The requirement on j_2 insures that no equation will be implicit in any of the remainder terms to be introduced, because the only possibility for an operator $:j_i:$ to leave the total rank of its argument unchanged is if the δ-rank is altered, *i.e.*, when $i = 1$. A consequence of these assumptions is that $\tau(k_i) = \tau(j_i)$.

We need keep only terms with a specific number of brackets in the expressions for F_i. If the smallest total rank of the j, $t \stackrel{\text{def}}{=} \min_i\{\tau(j_i)\}$, is at least 3, then the necessity of retaining each term is governed by the total rank criterion. One then keeps only terms of l brackets where

$$l(t-2) + t \leq N. \tag{3.10}$$

On the other hand, if the smallest total rank is 2, the necessity is governed separately by the total rank criterion ($N-3$ brackets) and the δ-rank from applying the $:j_1:$ ($N-1$ brackets), so the condition is

$$l \leq 2N - 4. \tag{3.11}$$

The inversion is accomplished iteratively on the total rank. The first step is to assume that k_i is j_i plus a remainder term of the same total rank,

$$k_i \to j_i + r_i^{(\tau(j_i))} \tag{3.12}$$

where r is a currently-unknown remainder term indexed by the δ-rank and the total rank. Formally, we make this substitution in the expressions (3.9); by the first assumption above about the F_i, both sides of each equation will have a term j_i, which may be eliminated. In the subsequent steps of the iteration, we may solve for the remainder term in the following way. Make a substitution for $r_i^{(t)}$ such that all terms of total rank t are cancelled off by setting it equal to the negative of all the other terms of that total rank in the expression, plus a remainder term of the next higher total rank;

$$r_i^{(t)} \to -W_i^{(t)} + r_i^{(t+1)}, \tag{3.13}$$

where $W_i^{(t)}$ consists of all terms of total rank t except j_i and the remainder term. Thus all terms of total rank t have been eliminated, and there are only terms of total rank $t+1$ and higher. We continue this iteration until the only terms that remain exceed the total rank cut-off N.

We now may find the answer by going back to the original substitution (3.12) and making the successive replacements (3.13); this gives the answer

$$k_i = j_i + \sum_{t=\tau(j_i)}^{N} -W_i^{(t)}. \tag{3.14}$$

It is tempting say that this can be obtained immediately without iteration, but keep in mind that W is not known at the current step until the last step has been solved; in general, that last step will have introduced new terms of the current total rank.

As it happens, the first remainder solution $r_i^{(i)}$ will yield the trivial answer $W_i^{(i)} = 0$ unless $\epsilon = 0$ for at least two non-zero terms j_m, j_n, with $m, n \geq 3$. This result is obtained in the exchange-and-seperate method for moving the first-order term (Section 4.4.2), where all j at each step have the same total rank. On the other hand, the solution of the subsequent remainders $r_i^{(t)}, t > i$, will yield the trivial answer $W_i^{(t)} = 0$ unless $\epsilon \geq 1$ for some j. This is obtained in moving higher-order terms (Section 4.4.4).

The solution to (3.8) may be obtained once and for all for a given N using this method by assuming the lowest possible total ranks on each of the j. Then at any step, one may use this result, discarding the terms whose total rank exceeds N or which involve a j that is zero. This has been done for $\tau(j) \geq 3$ and $N = 6$; the result is given in Appendix 4.C. Examples of the computation itself for less general cases are given in Sections 4.4.2 and 4.4.4.

4.4 Computation method for the concatenation formula

Selective use of the tools given will now allow us to acheive our goal of solving (1.11). The general idea is to move to the left all terms of lower δ-rank than their immediate left-hand neighbor. By "move to the left," we really mean to rewrite a pair of transformations in the standard factorization:

$$e^{:a_n:}e^{:b_m:} = e^{:k_1:}e^{:k_2:}e^{:k_3:}\dots e^{:k_N:}, \tag{4.1}$$

where $n \geq m$. Initially on the right side of (1.11) there is only one pair of transformations that satisfies this condition, $e^{:f_N:}e^{:g_1:}$. Once this is put into the standard factorization, other wrongly-ordered transformation pairs will occur, and these must be treated in a similar fashion, which will in turn give rise to more such pairs and so on.

The overall plan of attack is organized as follows. Initially, the first-order transformation will be moved to the left. This will leave behind many second and higher-order transformations, not at all in a suitable order. However, all the first-order transformations will be at the extreme left except for the intervening $e^{:f_2:}$, and they may be combined in a simple fashion. Now, the second-order transformations may moved to the left and $e^{:f_2:}$ to the right, leaving third and higher order transformations behind. Then the third-order ones may be moved; but, when they get to the left they must be combined into a single exponent and then this exponent split up by order. Fourth and higher orders are treated in a similar way. When this has been done through order $N-1$, the Lie transformations will be in the proper order.

The basic methods for treating pairs of transformations (4.1) are presented in the next Section, and the actual procedure for each order m in the succeeding Sections. As an alternative to the pairwise treatment of terms described above and elaborated below, one may try to combine all the terms of the concatenation problem (1.11) except $e^{:f_2:}$ and $e^{:g_2:}$ at once using an expanded version of the combine-and-separate method described in the next Section. This has a certain conceptual simplicity, but it is not practical as it would generate an inordinate number of terms for even small N.

4.4.1 Putting a pair of transformations into standard factorization

For the task of putting two successive transformations into standard factorization (4.1), there are three possibilities. If $m = 2$ and $\tau(b_2) = 2$, there is no choice but to use the exchange method. If this is not the case, there are two choices.

The *exchange-and-separate* method uses the exchange rule to write

$$e^{:a_n:}e^{:b_m:} = e^{:b_m:}e^{:e^{:-b_m:}a_n:} = e^{:b_m:}e^{:a_n + :-b_m:a_n + \cdots:}, \tag{4.2}$$

and then application of the separation procedure on the second transformation. Because the transformation $e^{:b_m:}$ remains separate on the left, this method is not useful for the case $m = n$. It is advantageous for $m = 1$ because all terms in the second transformation have total rank at least 3, reducing the number of Poisson brackets required.

The *combine-and-separate* method uses the BCH formula to combine the terms into a single exponent,

$$e^{:a_n:}e^{:b_m:} = e^{:a_n + b_m + \frac{1}{2}[a_n, b_m] + \cdots:}, \tag{4.3}$$

and then application of the separation procedure on this transformation. This method is preferable when $m > 1$ and necessary when $m = n$. Note that this incarnation of the BCH formula is distinct from that of the separation procedure that produces the functions F_i in (3.9); this is the combination of only two terms and changes at each step.

4.4.2 Moving the first-order term

We concentrate initially on the terms f_3 through g_1 of (1.11) and temporarily ignore the other Lie transformations. That is, we wish to put

$$e^{:f_1:}e^{:f_2:}e^{:f_3:}\dots e^{:f_N:}e^{:g_1:} \tag{4.4}$$

into standard factorization. This is done successively on each transformation; we start by putting the right-hand pair with f_N and g_1 into standard factorization. Then, with a first-order term on the left, we put the next pair with f_{N-1} into standard factorization, and so on until all the first-order terms are to the left of all terms of third order and higher. At each step, we have a pair of terms involving an f_n and some first-order term that will be rewritten in the standard factorization,

$$e^{:f_n:}e^{:b_1:} = e^{:k_1:}e^{:k_2:}e^{:k_3:}\ldots. \tag{4.5}$$

Although it is possible to use the combine-and-separate method instead of the exchange-and-separate, it is not practical. With $\tau(b_1) = 2$, there are many more terms in the BCH series that need to be retained, because no total rank 2 term occurs in the latter method; it is exactly the condition (3.11) versus (3.10). For large N, there are an enormous number of terms, so it is more practical to use the exchange-and-separate method for the first-order transformations.

In order to solve (4.5) for $n \geq 3$, we first make the first-order term specific and use the exchange rule to write

$$\begin{aligned} e^{:f_n:}e^{:b_1:} &= e^{:b_1:}e^{:e^{:-b_1:}f_n:} \\ &= e^{:b_1:}e^{:f_n + [-b_1, f_n] + \cdots + :-b_1:^{n-1} f_n/(n-1)!:}. \end{aligned} \tag{4.6}$$

Note that this second transformation has only a finite number of terms in it, and of each order from n down through 1, because of the order-reducing property of a first-order Lie operator. Rewriting each of these terms as $j_{n-i} \stackrel{\text{def}}{=} \frac{1}{i!}:-b_1:^i f_n$ for $0 \leq i \leq n-1$, we have

$$e^{:f_n:}e^{:b_1:} = e^{:b_1:}e^{:j_1 + j_2 + \cdots + j_n:}. \tag{4.7}$$

We may now apply the separation procedure to put the right-hand side of this equation into the standard factorization (3.8), which we can then substitute into (4.7) to get

$$e^{:f_n:}e^{:b_1:} = e^{:b_1:}e^{:k_1:}e^{:k_2:}\ldots e^{:k_m:}. \tag{4.8}$$

The next step is to combine the two adjacent first-order transformations into a single exponent. This is a trivial application of the BCH formula; because they are first-order terms, their Poisson brackets are constants and may therefore be discarded. Thus we simply put the sum of the two first-order terms into a single exponent:

$$e^{:f_n:}e^{:b_1:} = e^{:b_1 + k_1:}e^{:k_2:}\ldots e^{:k_m:}. \tag{4.9}$$

Each of the Lie transformation pairs that is put into standard factorization according to (4.5) produces a set of polynomials $\{k_i^{(t)}\}$ which we label with a superscript that superficially indicates the total rank. If $i \leq t$, it is the total rank. If $i > t$, the total rank is at least i, but the term was produced in the same step as the total rank t terms. At the first step involving f_N, all the polynomials produced will be of total rank N, at the next they will be of total rank $N-1$, and at the step i mostly of total rank $N-i+1$ but some may be higher.

The first-order terms are accumulated at each step according to (4.9). On the first step, $b_1 = g_1$, and at each step the particular k_1 is added. Thus the final first-order term is g_1 plus the sum of all the k_1 terms,

$$G_1 \stackrel{\text{def}}{=} g_1 + \sum_t k_1^{(t)}. \tag{4.10}$$

When all the terms of the type (4.5) in (4.4) have been treated in succession we will end up with the result

$$\begin{aligned} e^{:f_1:}e^{:f_2:}e^{:f_3:}\dots e^{:f_N:}e^{:g_1:} &= e^{:f_1:}e^{:f_2:}e^{:G_1:} \\ &\times e^{:k_2^{(3)}:}e^{:k_3^{(3)}:}\dots e^{:k_{N-1}^{(3)}:}\dots \\ &\times e^{:k_2^{(N-1)}:}\dots e^{:k_{N-1}^{(N-1)}:}e^{:k_2^{(N)}:}\dots \\ &\times e^{:k_{N-1}^{(N)}:}e^{:k_N^{(N)}:}, \end{aligned} \tag{4.11}$$

with the first-order terms to the left of all third- and higher-order terms. It should be noted that there will be no $k_N^{(l)}$ term for $l \leq N-1$ because such a term can be formed only by taking nested Poisson brackets solely of $j_l^{(l)}$, up to the maximum number of brackets that the total rank criterion will allow; the Poisson bracket $[j_l^{(l)}, j_l^{(l)}]$ though, is zero.

In order to put the right-hand side of (4.11) into standard factorization, we need to move $e^{:f_2:}$ to the right of $e^{:G_1:}$. For this we just use the exchange rule as described in the next Section, and then the two adjacent first-order transformations may be combined by adding their exponents. The higher-order terms may be put into standard factorization by the procedure described in Section 4.4.4. Because the second-order terms $e^{:k_2^{(t)}:}$ have total rank higher than two, it is possible to move and combine them using the rules for higher-order transformations instead of using the exchange rule.

To see how this works in practice, let us look at a simple example. Suppose we take $N = 4$, and consider the part of the problem where we will move the first-order term to the left of the third-order term. That is, we will write

$$e^{:f_3:}e^{:g_1:} = e^{:k_1:}e^{:k_2:}e^{:k_3:}e^{:k_4:}, \tag{4.12}$$

and determine the polynomials k_n. Following the procedure above, we write

$$e^{:f_3:}e^{:g_1:} = e^{:g_1:}e^{:j_1^{(3)} + j_2^{(3)} + j_3^{(3)}:}, \tag{4.13}$$

with the $j_n^{(3)}$ given by the exchange rule,

$$\begin{aligned} j_1^{(3)} &= \tfrac{1}{2}[g_1,[g_1,f_3]] \\ j_2^{(3)} &= -[g_1,f_3] \\ j_3^{(3)} &= f_3. \end{aligned} \qquad (4.14)$$

We now break apart the exponent into separate transformations. The answer may be obtained from Appendix 4.C, or we may perform the separation procedure ourselves. Since each of these terms is total rank 3 and we are keeping through $N = 4$, we will need the BCH formula (3.7) only through the one-Poisson-bracket term, that is, equation (3.9) is explicitly in this case,

$$\begin{aligned} j_1^{(3)} &= F_1(k_1^{(3)},k_2^{(3)},k_3^{(3)},k_4^{(3)}) = k_1^{(3)} + \tfrac{1}{2}[k_1^{(3)},k_2^{(3)}] \\ j_2^{(3)} &= F_2(k_1^{(3)},k_2^{(3)},k_3^{(3)},k_4^{(3)}) = k_2^{(3)} + \tfrac{1}{2}[k_1^{(3)},k_3^{(3)}] \\ j_3^{(3)} &= F_3(k_1^{(3)},k_2^{(3)},k_3^{(3)},k_4^{(3)}) = k_3^{(3)} + \tfrac{1}{2}[k_2^{(3)},k_3^{(3)}] + \tfrac{1}{2}[k_1^{(3)},k_4^{(3)}] \\ 0 &= F_4(k_1^{(3)},k_2^{(3)},k_3^{(3)},k_4^{(3)}) = k_4^{(3)} + \tfrac{1}{2}[k_2^{(3)},k_4^{(3)}]. \end{aligned} \qquad (4.15)$$

Making the initial substitution $k_i^{(3)} \to j_i^{(3)} + r_i^{(3)}$ (3.12) and then the next substitution $r_i^{(3)} \to -W_i^{(3)} + r_i^{(4)}$ (3.13) yields the result $W^{(3)} = 0$ (because all Poisson bracket terms are of higher order than 3) and the residual equations

$$\begin{aligned} 0 &= r_1^{(4)} + \tfrac{1}{2}[j_1^{(3)},j_2^{(3)}] + \tfrac{1}{2}[r_1^{(4)},j_2^{(3)}] + \tfrac{1}{2}[j_1^{(3)},r_2^{(4)}] + \tfrac{1}{2}[r_1^{(4)},r_2^{(4)}] \\ 0 &= r_2^{(4)} + \tfrac{1}{2}[j_1^{(3)},j_3^{(3)}] + \tfrac{1}{2}[r_1^{(4)},j_3^{(3)}] + \tfrac{1}{2}[j_1^{(3)},r_3^{(4)}] + \tfrac{1}{2}[r_1^{(4)},r_3^{(4)}] \\ 0 &= r_3^{(4)} + \tfrac{1}{2}[j_2^{(3)},j_3^{(3)}] + \tfrac{1}{2}[r_2^{(4)},j_3^{(3)}] + \tfrac{1}{2}[j_2^{(3)},r_3^{(4)}] + \tfrac{1}{2}[r_2^{(4)},r_3^{(4)}] \\ &\quad + \tfrac{1}{2}[j_1^{(3)},r_4^{(4)}] + \tfrac{1}{2}[r_1^{(4)},r_4^{(4)}] \\ 0 &= r_4^{(4)} + \tfrac{1}{2}[j_2^{(3)},r_4^{(4)}] + \tfrac{1}{2}[r_2^{(4)},r_4^{(4)}], \end{aligned} \qquad (4.16)$$

after cancelling off the $j_n^{(3)}$. We have made the substitution $k_4^{(3)} \to r_4^{(4)}$ because there is no $j_4^{(3)}$.

Looking only at terms of total rank 4, and solving for $r_i^{(4)}$ according to (3.13), we find

$$\begin{aligned} r_1^{(4)} &= -\tfrac{1}{2}[j_1^{(3)},j_2^{(3)}] + r_1^{(5)} \\ r_2^{(4)} &= -\tfrac{1}{2}[j_1^{(3)},j_3^{(3)}] + r_2^{(5)} \\ r_3^{(4)} &= -\tfrac{1}{2}[j_2^{(3)},j_3^{(3)}] + r_3^{(5)} \\ r_4^{(4)} &= 0. \end{aligned} \qquad (4.17)$$

The new remainder term and the three remaining Poisson brackets that occur in each of the equations (4.16) are of total rank at least 5. Since our truncation criterion is to drop terms total rank 5 and greater, we may discard them. We are thus at the end of the iteration, and we find the answer by adding up all the remainders according to (3.14). There was only one remainder in this example, so

$$\begin{aligned} k_1^{(3)} &= j_1^{(3)} - \tfrac{1}{2}[j_1^{(3)}, j_2^{(3)}] \\ k_2^{(3)} &= j_2^{(3)} - \tfrac{1}{2}[j_1^{(3)}, j_3^{(3)}] \\ k_3^{(3)} &= j_3^{(3)} - \tfrac{1}{2}[j_2^{(3)}, j_3^{(3)}] \\ k_4^{(3)} &= 0. \end{aligned} \tag{4.18}$$

Substituting back into (4.13), we conclude

$$\begin{aligned} e^{:f_3:}e^{:g_1:} &= e^{:g_1:}e^{:j_1^{(3)} - \frac{1}{2}[j_1^{(3)}, j_2^{(3)}]:} \\ &\quad \times e^{:j_2^{(3)} - \frac{1}{2}[j_1^{(3)}, j_3^{(3)}]:}e^{:j_3^{(3)} - \frac{1}{2}[j_2^{(3)}, j_3^{(3)}]:} \\ &= e^{:g_1:}e^{:\frac{1}{2}[g_1,[g_1,f_3]] - \frac{1}{2}[\frac{1}{2}[g_1,[g_1,f_3]], -[g_1,f_3]]:} \\ &\quad \times e^{:-[g_1,f_3] - \frac{1}{2}[\frac{1}{2}[g_1,[g_1,f_3]], f_3]:}e^{:f_3 - \frac{1}{2}[-[g_1,f_3], f_3]:}. \end{aligned} \tag{4.19}$$

It is then possible to simplify the nested Poisson brackets.

This is a very simple example, and it may seem that it is easily solved intuitively, without the formalism. This is so, but for higher orders the formalism is necessary. Appendix 4.A describes some of the Lie algebraic manipulation that can keep the calculation under control. This example is unfortunately too simple to show that terms $k_i^{(n)}$ with $i > n$ do occur, which only happens for $N \geq 5$.

4.4.3 Moving second-order terms

Moving the second-order terms in (4.11) to the right of the first-order terms or moving them to the left of the higher-order terms is a pure application of the exchange rule. Since a second-order transformation is just a linear transformation on phase space, one can effectively rewrite a pair of Lie transformations involving a second-order transformation as

$$e^{:a(\zeta):}e^{:b_2:} = e^{:b_2:}e^{:e^{:-b_2:}a(\zeta):} = e^{:b_2:}e^{:a(e^{:-b_2:}\zeta):} \tag{4.20}$$

Therefore, one would move the second-order terms by transforming the intervening operators' arguments. For instance, the first three terms on the right-hand side of (4.11) are rewritten

$$e^{:f_1:}e^{:e^{:f_2:}g_1:}e^{:f_2:} = e^{:f_1 + e^{:f_2:}g_1:}e^{:f_2:} = e^{:f_1 + g_1^T:}e^{:f_2:}, \tag{4.21}$$

where here g_1^T is defined as the transformed polynomial

$$g_1^T(\zeta) = e^{:f_2:} g_1(\zeta) = g_1(e^{:f_2:}\zeta). \tag{4.22}$$

By so moving all second-order transformations to the right of first-order transformations and to the left of all higher-order ones, we will then be left with all second order terms together in the proper δ-rank sequence. From an aesthetic point of view, it is now desirable to combine these second-order terms into a single second-order term. From a practical point of view, however, this is not necessary because the linear transformations are all kept as matrices when doing numerical computations and matrix multiplication is computationally simple. This is a good thing, because it is not possible in general to combine all these terms. There are two terms, $e^{:f_2:}$ and $e^{:g_2:}$, that are of total rank 2, which means that any BCH series involving it does not terminate. On the other hand, all the other second-order transformations are of total rank 3 and higher, so that application of the BCH formula with appropriate truncation according to the total rank criterion will yield an answer with a finite number of terms.

4.4.4 Moving higher-order terms

Moving transformations of order 3 and higher is essentially the same problem mathematically as that of the first-order terms, but because it is simpler, and because we need to combine terms of the same order, we use the combine-and-separate method rather than the exchange-and-separate method. Ignoring the first- and second-order terms, which now have been pushed to the left, we may look at the concatenation problem as reformulating a product of transformations each of third or higher order into the standard factorization. Specifically, we look at the problem as putting

$$e^{:f_3:}e^{:f_4:}\dots e^{:f_N:}\; e^{:g_3:}e^{:g_4:}\dots e^{:g_N:} \tag{4.23}$$

into standard factorization. Although there are actually many terms of each order 3 and higher resulting from the first-order calculation, knowing this solution will allow us to apply it repeatedly to obtain the general solution. As with moving the first-order term, we start by moving the term $e^{:g_3:}$ leftward, ignoring all the other g transformations, then proceed to move the term $e^{:g_4:}$, and so on. This process gets successively easier.

The elemental task in this process is to re-express the two adjacent terms $e^{:a_n:}e^{:b_m:}$ in the standard factorization. Because now $n \geq 3$ and $m \geq 3$, there is no problem with first- or second-order terms recurring. Using the combine-and-separate method, the BCH formula is applied to obtain

$$e^{:a_n:}e^{:b_m:} = e^{:a_n + b_m + \frac{1}{2}[a_n, b_m] + \cdots :}. \tag{4.24}$$

The combined terms can be any δ-rank from m through N, although by the nature of the BCH formula, only certain ones will appear. After applying the separation procedure, this exponent will be expressed as a product of transformations in the standard factorization. In this case, we are solving equation (3.8), but with no first- or second-order terms, and perhaps some other terms missing as well.

After a transformation of a given-order is moved to the left, it must be combined with the other term of the same δ-rank. In this respect, the problem is very different from the first-order case, because there the combination of the terms of like rank was trivial. Here, the combination once again produces a whole menagerie of terms and must be treated in the same way as when any other term is encountered.

Let us consider specifically one of these elemental tasks with $N = 6$, in fact one of combining terms of like δ-rank, that of finding the standard factorization for $e^{:f_3:}e^{:g_3:}$. Using the BCH formula, we write

$$\begin{aligned} e^{:f_3:}e^{:g_3:} &= \exp\big(f_3 + g_3 + \tfrac{1}{2}[f_3, g_3] + \tfrac{1}{12}[f_3, [f_3, g_3]] \\ &\qquad + \tfrac{1}{12}[g_3, [g_3, f_3]] - \tfrac{1}{24}[f_3, [g_3, [f_3, g_3]]]\big) \\ &= e^{:k_3:}e^{:k_4:}e^{:k_5:}e^{:k_6:}, \end{aligned} \tag{4.25}$$

and solve for k. Order by order, the terms combined in the exponent are

$$\begin{aligned} j_3 &= f_3 + g_3 \\ j_4 &= \tfrac{1}{2}[f_3, g_3] \\ j_5 &= \tfrac{1}{12}[f_3, [f_3, g_3]] + \tfrac{1}{12}[g_3, [g_3, f_3]] \\ j_6 &= -\tfrac{1}{24}[f_3, [g_3, [f_3, g_3]]]. \end{aligned} \tag{4.26}$$

We now apply the separation procedure. As in Section 4.4.2, we may consult Appendix 4.C for the general solution and apply it to this case, or we may perform the separation procedure ourselves. Doing the latter, we use the BCH formula to find the functions F_i of (3.9),

$$\begin{aligned} &e^{:k_3:}e^{:k_4:}e^{:k_5:}e^{:k_6:} \\ &= e^{:k_3 + k_4 + k_5 + k_6 + \frac{1}{2}[k_3, k_4] + \frac{1}{2}[k_3, k_5] + \frac{1}{12}[k_3, [k_3, k_4]]:}, \end{aligned} \tag{4.27}$$

$$\begin{aligned} j_3 &= F_3(k_3, k_4, k_5, k_6) = k_3 \\ j_4 &= F_4(k_3, k_4, k_5, k_6) = k_4 \\ j_5 &= F_5(k_3, k_4, k_5, k_6) = k_5 + \tfrac{1}{2}[k_3, k_4] \\ j_6 &= F_6(k_3, k_4, k_5, k_6) = k_6 + \tfrac{1}{2}[k_3, k_5] + \tfrac{1}{12}[k_3, [k_3, k_4]]. \end{aligned} \tag{4.28}$$

We make the substitutions

$$\begin{aligned} k_3 &\to j_3 + r_3^{(3)} \\ k_4 &\to j_4 + r_4^{(4)} \\ k_5 &\to j_5 + r_5^{(5)} \\ k_6 &\to j_6 + r_6^{(6)}. \end{aligned} \tag{4.29}$$

and get, after cancelling off the j terms,

$$\begin{aligned}
0 &= r_3^{(3)} \\
0 &= r_4^{(4)} \\
0 &= r_5^{(5)} + \tfrac{1}{2}[j_3 + r_3^{(3)}, j_4 + r_4^{(4)}] \\
0 &= r_6^{(6)} + \tfrac{1}{2}[j_3 + r_3^{(3)}, j_5 + r_5^{(5)}] \\
&\quad + \tfrac{1}{12}[j_3 + r_3^{(3)}, [j_3 + r_3^{(3)}, j_4 + r_4^{(4)}]].
\end{aligned} \tag{4.30}$$

Solving for the r and substituting in (4.29) gives

$$\begin{aligned}
k_3 &= j_3 = f_3 + g_3 \\
k_4 &= j_4 = \tfrac{1}{2}[f_3, g_3] \\
k_5 &= j_5 - \tfrac{1}{2}[j_3, j_4] \\
&= \tfrac{1}{12}[f_3, [f_3, g_3]] + \tfrac{1}{12}[g_3, [g_3, f_3]] - \tfrac{1}{2}[f_3 + g_3, \tfrac{1}{2}[f_3, g_3]] \\
k_6 &= j_6 - \tfrac{1}{2}[j_3, j_5] - \tfrac{1}{12}[j_3, [j_3, j_4]] \\
&= -\tfrac{1}{24}[f_3, [g_3, [f_3, g_3]]] - \tfrac{1}{2}[f_3 + g_3, \tfrac{1}{12}[f_3, [f_3, g_3]] \\
&\quad + \tfrac{1}{12}[g_3, [g_3, f_3]]] - \tfrac{1}{12}[f_3 + g_3, [f_3 + g_3, \tfrac{1}{2}[f_3, g_3]].
\end{aligned} \tag{4.31}$$

This is already the complete solution; there is no further need to iterate because the total rank of each term equals the δ-rank.

4.5 Summary

The transfer map arising from a Hamiltonian flow may be represented, including inhomogeneities, through a given order by a product of Lie transformations on phase space factored by order. Two such maps may be combined into a single map using the concatenation formula whose derivation is outlined above. In order to calculate the concatenation formula and the Lie products, it is necessary that the inhomogeneity be small. How this smallness relates to the smallness of the phase space perturbation is imposed in a general way by the algebra; one is then free to choose specifically the criterion to be in accord with physical considerations. Here, the criterion choosen is that these two quantities count equally and no term beyond a certain order N is retained.

The concatenation formula is accurate through the order computed without approximation, including feed-down effects from the inhomogeneity. It will not produce the same result as when the two transformations are computed in succession because of effects beyond order N that are lost in the concatenation. Thus if these effects are important concatenation is not suitable.

The concatenation formula have been computed and checked through $N = 6$. The result and the details of the symbolic manipulation used are given in the Appendices.

4.A A basis for the Lie algebra

The computations described above involve a considerable amount of algebraic manipulation, particularly for computations of higher N. For $N \gtrsim 5$ it becomes unfeasible to do these computations by hand, and more practical to automate this process. There are several stages where it is useful to use symbolic computation. First of all, the BCH formula (3.7) may be obtained from references (for example [15]) for the first few terms, but for more terms, it may be necessary to compute them. This can be done symbolically using a recursive procedure derived from the formula in [9], or by other methods. Second, the separation procedure may be implemented automatically to take any set of functions F_i (3.9), and invert them. Finally, throughout all these computations it is vital, particularly for large N, to keep the expressions to a manageable size. This is done by reducing the terms to a minimal independent set, *i.e.*, a basis. The last two tasks have been implemented as a program in the symbolic manipulation program *SMP* [13]. Interested readers should contact the first author for details.

In this Appendix we present the last of these tasks; a basis for the Lie algebra of up to four quantities, some of which may be duplicated. A reduction to a basis is possible because of the axioms of a Lie algebra (1.5–1.7). This builds on the work of Dragt and Forest [5], who presented a basis for an arbitrary number of distinct quantities, but the orientation here is algorithmic, *i.e.*, the way a symbolic manipulation package would project a term onto the basis.

To aid in this effort, we shall need some notation and terminology. The term *atom* will mean a single symbol that represents something in the Lie algebra. A *bracket* is a less fundamental quantity, being a bracket in the Lie algebra of either atoms or brackets. Everything in the Lie algebra is either an atom or a bracket, or both. Since we are dealing with formal symbol manipulation, the distinction between atoms and brackets is arbitrary and has no mathematical meaning. In fact, it will prove convenient to temporarily manipulate some brackets as atoms. Since they are just symbols, atoms will be thought of as having no other property than their lexical order with respect to other atoms. When a bracket is to be thought of as an atom, its lexical position will be given explicitly. The symbols 'a', 'b', 'c' and 'd' used in this Appendix are not the elements of the Lie algebra themselves, but stand for these elements considered as symbols, either atoms or brackets. For example, $[a, b]$ may be lexically ordered or not depending on what symbols a and b stand for.

A *nest* is a special kind of bracket that contains an atom in the first position and a nest or an atom in the second. We shall denote a nest by leaving off all but the outer pair of brackets, *e.g.*, $[f, g, f, g]$ means $[f, [g, [f, g]]]$. Second, we shall classify the repetition of atoms in a term by numerical counts of each different one. For example, $[f, g, f, g]$ is in the class 2×2 and $[f, g, f, h]$ is in the class $2 \times 1 \times 1$. This terminology is not restricted to nests.

Let us see how to rewrite a term in the basis for n different atoms, or $1 \times 1 \times \cdots \times 1$, as derived from Dragt and Forest [5]. Any member of the Lie algebra of this type may be reduced to a nest with the lexically first (or any particular) atom last, as may be shown by induction. First, suppose that the number of atoms is two. Then the term is automatically a nest. If the lexically first atom is not last, we use antisymmetry to rewrite the term so that it is last.

Suppose we have a term of brackets that is not a nest; $[a_1, a_2, \ldots, a_n]$, where any term except the last may be some bracket. Let us look at one specifically, and say that $a_i = [c, d]$. Then the term is $[a_1, \ldots, a_{i-1}, [c, d], a_{i+1}, \ldots, a_n]$. Using the Jacobi identity, we may rewrite this as

$$\begin{aligned} [a_1, a_2, \ldots, a_n] &= [a_1, \ldots, a_{i-1}, c, d, a_{i+1}, \ldots, a_n] \\ &\quad -[a_1, \ldots, a_{i-1}, d, c, a_{i+1}, \ldots, a_n]. \end{aligned} \tag{A.1}$$

Now each of c and d may themselves be brackets, but they will have fewer total atoms than $[c, d]$, so if we apply this transformation repeatedly, we will eventually have only single atoms, and therefore a nest.

If the lexically first atom is not last in this nest, we must rewrite the nest so that it is. Let us suppose in $[a_1, a_2, \ldots, a_n]$ that each of the a_j stands for an atom and that a_i is the lexically first atom. If $i = n - 1$, then we apply antisymmetry and we are done. If $i < n - 1$, call $b \stackrel{\text{def}}{=} [a_{i+2}, a_{i+3}, \ldots, a_n]$. Consider now the term written supposing that b stands for an atom and with the last two brackets explicit; then apply the Jacobi identity and antisymmetry:

$$\begin{aligned} [a_1, \ldots, a_n] &= [a_1, \ldots, a_{i-1}, a_i, a_{i+1}, b] = [a_1, \ldots, a_{i-1}, [a_i, [a_{i+1}, b]]] \\ &= -[a_1, \ldots, a_{i-1}, [a_{i+1}, [b, a_i]]] + [a_1, \ldots, a_{i-1}, [b, [a_{i+1}, a_i]]] \\ &= -[a_1, \ldots, a_{i-1}, a_{i+1}, b, a_i] + [a_1, \ldots, a_{i-1}, b, a_{i+1}, a_i]. \end{aligned} \tag{A.2}$$

Each of these terms has the a_i in the final position. Of course they may not be nests, because b can be a bracket, but by reapplying the nesting algorithm (A.1), which does not alter the position of the last atom, we will have expressed the original term as the sum of nests that have the lexically first atom last.

We now need to consider the case of duplicate atoms. Of course, all the reductions made for the case of distinct atoms apply here, but there will be further reductions that can be made if some of the quantities are the same. The general technique is to start with the Jacobi identity on the outer nest, then successively apply antisymmetry and the inner Jacobi identities to find identities. Unfortunately, a general treatment for arbitrary numbers of atoms is more involved than in the case of distinct atoms. Here, we will consider the particular terms that arise in the calculation for $N = 6$. Different aspects of the general problem are addressed in more detail in [1].

For the calculation through total order $N = 6$, no term of more than four atoms will occur. Therefore, let us consider the possibilities for three and four atoms. For three atoms, the only case is 2×1. There is clearly only one term: $[f, g, f]$ if it is the lexically first term that is duplicated, or $[g, g, f]$ if not.

For four atoms, there are three possible situations, $2 \times 1 \times 1$, 3×1, and 2×2. There are respectively three, one, and one term in the bases for each. To analyze these three cases, one should start with the six terms arising when all atoms are different:

$$[k, h, g, f], \quad [k, g, h, f], \quad [h, k, g, f], \quad [h, g, k, f], \quad [g, h, k, f], \quad [g, k, h, f]. \tag{A.3}$$

Now we may identify k with h to find three different terms:

$$[h, h, g, f], \quad [h, g, h, f], \quad [g, h, h, f]. \tag{A.4}$$

For duplication of the lexically first atom, one must make use of the Jacobi identity on the outer bracket, to reduce the number of terms from four to three:

$$[k, f, h, f], \quad [h, f, k, f], \quad [f, k, h, f]. \tag{A.5}$$

We see easily that for 3×1 there is one term: $[g, g, g, f]$ (or $[f, f, g, f]$).

Now consider the 2×2 case. At first glance, it looks like there are two terms in the basis, $[g, f, g, f]$ and $[f, g, g, f]$. However, these can be shown to be equal by applying the Jacobi identity on the outer brackets, and antisymmetry:

$$\begin{aligned} [f, g, g, f] + [g, [g, f], f] + [[g, f], f, g] &= 0 \\ [f, g, g, f] - [g, f, g, f] &= 0. \end{aligned} \tag{A.6}$$

We have now shown that any member of the Lie algebra of atoms can be reexpressed as the linear combination of nests of these atoms with the lexically first atom last. Accounting for duplicate atoms, this set can be made smaller to form a basis. This is not necessarily the representation in the fewest number of terms, nor is it the most natural way of expressing the equations for efficient numerical computation. For either of these alternate goals, it is not obvious which way to express the equations. In fact one must try many possiblities, guided by intuition, to improve the expression.

4.B Concatenation formulae for $N = 6$

In this Appendix, the results for the computation described above for $N = 6$ are presented for reference.

First, we consider the problem of concatenating just the first-order term. The tables 4.1-4.3 at the end of this Appendix give the solution for the terms on the right side of (4.11). In these tables, the k are described in terms of j. Since these j always have the same superscript as the k, it has been left off. Note in the tables the manifestation of the condition determining the number of brackets required (3.10): for the $k^{(3)}$ expressions, there are at most 3 brackets in a term, for the $k^{(4)}$ calculations, at most 1, and for $k^{(5)}$ and $k^{(6)}$, no brackets. The expressions for k may be obtained from Appendix 4.C by keeping only these number of brackets and setting the appropriate j to zero.

As mentioned in Section 4.4.2, it is possible to apply the concatenation rules for second- and higher-order to rewrite (4.11) in the standard factorization. In this case,

$$e^{:f_1:}e^{:f_2:}e^{:f_3:}e^{:f_4:}e^{:f_5:}e^{:f_6:}e^{:g_1:} = e^{:h_1:}e^{:f_2:}e^{:h_2:}e^{:h_3:}e^{:h_4:}e^{:h_5:}e^{:h_6:} \tag{B.1}$$

where

$$\begin{aligned}
h_1 &= f_1 + e^{:f_2:}(g_1 + k_1^{(3)} + k_1^{(4)} + k_1^{(5)} + k_1^{(6)}) \\
h_2 &= k_2^{(3)} + k_2^{(4)} + k_2^{(5)} + k_2^{(6)} + \tfrac{1}{2}[k_2^{(3)}, k_2^{(4)} + k_2^{(5)}] + \tfrac{1}{12}[k_2^{(3)}, k_2^{(3)}, k_2^{(4)}] \\
h_3 &= k_3^{(3)} + k_3^{(4)} + k_3^{(5)} + k_3^{(6)} - [k_2^{(4)} + k_2^{(5)}, k_3^{(3)}] \\
h_4 &= k_4^{(3)} + k_4^{(4)} + k_4^{(5)} + k_4^{(6)} + \tfrac{1}{2}[k_3^{(3)}, k_3^{(4)} + k_3^{(5)}] \\
h_5 &= k_5^{(3)} + k_5^{(4)} + k_5^{(5)} + k_5^{(6)} - \tfrac{1}{6}[k_3^{(3)}, k_3^{(3)}, k_3^{(4)}] \\
h_6 &= k_6^{(6)}.
\end{aligned} \tag{B.2}$$

The formula for concatenation of the higher order terms, as described in Section 4.4.4, is fortunately much simpler. We know first of all that second-order transformations are handled using the exchange rule. We may thus consider the problem to be

$$e^{:f_3:}e^{:f_4:}e^{:f_5:}e^{:f_6:}\, e^{:g_3:}e^{:g_4:}e^{:g_5:}e^{:g_6:} = e^{:h_3:}e^{:h_4:}e^{:h_5:}e^{:h_6:}. \tag{B.3}$$

The solution is

$$\begin{aligned}
h_3 &= f_3 + g_3 \\
h_4 &= f_4 + g_4 + \tfrac{1}{2}[f_3, g_3] \\
h_5 &= f_5 + g_5 - [g_3, f_4] - \tfrac{1}{6}:f_3:^2 g_3 + \tfrac{1}{3}:g_3:^2 f_3 \\
h_6 &= f_6 + g_6 - [g_3, f_5] + \tfrac{1}{2}:g_3:^2 f_4 + \tfrac{1}{2}[f_4, g_4] - \tfrac{1}{4}[f_4, f_3, g_3] \\
&\quad -\tfrac{1}{4}[g_4, f_3, g_3] + \tfrac{1}{24}:f_3:^3 g_3 - \tfrac{1}{8}:g_3:^3 f_3 + \tfrac{1}{8}[f_3, g_3, f_3, g_3].
\end{aligned} \tag{B.4}$$

Again, the reader is reminded that this is not necessarily the most compact or computationally efficient form. For purposes of numerical coding, it is possible to rearrange the terms so that quantities already calculated can be reused to maximum benefit. In such a code, there are three basic routines, one that moves the first-order transformation according to (B.2), one that does the linear transformations on the polynomials according to the exchange rule, and a third that moves the higher-order terms according to (B.4).

$k_1^{(3)}$	$\tau = 3$	$j_1 + \frac{1}{2}[j_2, j_1] - \frac{1}{6}[j_1, j_3, j_1] + \frac{1}{6}[j_2, j_2, j_1]$ $-\frac{1}{8}[j_1, j_3, j_2, j_1] - \frac{1}{24}[j_2, j_1, j_3, j_1] + \frac{1}{24}[j_2, j_2, j_2, j_1]$
$k_2^{(3)}$	$\tau = 3$	$j_2 + \frac{1}{2}[j_3, j_1] - \frac{1}{12}[j_2, j_3, j_1] + \frac{1}{6}[j_3, j_2, j_1]$ $-\frac{1}{24}[j_2, j_3, j_2, j_1] - \frac{1}{24}[j_3, j_1, j_3, j_1] + \frac{1}{24}[j_3, j_2, j_2, j_1]$
$k_3^{(3)}$	$\tau = 3$	$j_3 + \frac{1}{2}[j_3, j_2] - \frac{1}{6}[j_2, j_3, j_2] + \frac{1}{6}[j_3, j_3, j_1] + \frac{1}{24}[j_2, j_2, j_3, j_2]$ $-\frac{1}{8}[j_2, j_3, j_3, j_1] + \frac{1}{24}[j_3, j_2, j_3, j_1] + \frac{1}{24}[j_3, j_3, j_2, j_1]$
$k_4^{(3)}$	$\tau = 5$	$-\frac{1}{12}[j_3, j_3, j_2] + \frac{1}{24}[j_3, j_2, j_3, j_2] - \frac{1}{24}[j_3, j_3, j_3, j_1]$
$k_5^{(3)}$	$\tau = 6$	$\frac{1}{24}[j_3, j_3, j_3, j_2]$

$j_1^{(3)}$	$\frac{1}{2}{:}{-g_1}{:}^2 f_3$
$j_2^{(3)}$	${:}{-g_1}{:}\, f_3$
$j_3^{(3)}$	f_3

TABLE 4.1. The polynomials $k^{(3)}$.

$k_1^{(4)}$	$\tau = 4$	$j_1 + \frac{1}{2}[j_2, j_1]$
$k_2^{(4)}$	$\tau = 4$	$j_2 + \frac{1}{2}[j_3, j_1]$
$k_3^{(4)}$	$\tau = 4$	$j_3 + \frac{1}{2}[j_3, j_2] + \frac{1}{2}[j_4, j_1]$
$k_4^{(4)}$	$\tau = 4$	$j_4 + \frac{1}{2}[j_4, j_2]$
$k_5^{(4)}$	$\tau = 6$	$\frac{1}{2}[j_4, j_3]$

$j_1^{(4)}$	$\frac{1}{6}{:}{-g_1}{:}^3 f_4$
$j_2^{(4)}$	$\frac{1}{2}{:}{-g_1}{:}^2 f_4$
$j_3^{(4)}$	${:}{-g_1}{:}\, f_4$
$j_4^{(4)}$	f_4

TABLE 4.2. The polynomials $k^{(4)}$.

$k_1^{(5)} = j_1^{(5)}$	$\frac{1}{24}{:}{-g_1}{:}^4 f_5$
$k_2^{(5)} = j_2^{(5)}$	$\frac{1}{6}{:}{-g_1}{:}^3 f_5$
$k_3^{(5)} = j_3^{(5)}$	$\frac{1}{2}{:}{-g_1}{:}^2 f_5$
$k_4^{(5)} = j_4^{(5)}$	${:}{-g_1}{:}\, f_5$
$k_5^{(5)} = j_5^{(5)}$	f_5

$k_1^{(6)} = j_1^{(6)}$	$\frac{1}{120}{:}{-g_1}{:}^5 f_6$
$k_2^{(6)} = j_2^{(6)}$	$\frac{1}{24}{:}{-g_1}{:}^4 f_6$
$k_3^{(6)} = j_3^{(6)}$	$\frac{1}{6}{:}{-g_1}{:}^3 f_6$
$k_4^{(6)} = j_4^{(6)}$	$\frac{1}{2}{:}{-g_1}{:}^2 f_6$
$k_5^{(6)} = j_5^{(6)}$	${:}{-g_1}{:}\, f_6$
$k_6^{(6)} = j_6^{(6)}$	f_6

TABLE 4.3. The polynomials $k^{(5)}$ and $k^{(6)}$.

4.C The results of the separation procedure for $N = 6$

In this Appendix the solution to equation (3.8) is presented for all j terms of total rank at least 3 through a cutoff $N = 6$. This is sufficient to calculate all the results of Appendix 4.B using the exchange-and-separate method on the first-order transformations. By condition (3.10), we need at most 3 Poisson brackets in the BCH formula; this is as given explicitly in (3.7). Therefore, by combining the multiple exponents, we have

$$\begin{aligned}
j_1 &= k_1 - \tfrac{1}{2}[k_2, k_1] - \tfrac{1}{12}[k_1, k_3, k_1] + \tfrac{1}{12}[k_2, k_2, k_1] + \tfrac{1}{12}[k_1, k_3, k_2, k_1] \\
j_2 &= k_2 - \tfrac{1}{2}[k_3, k_1] - \tfrac{1}{12}[k_1, k_4, k_1] - \tfrac{1}{6}[k_2, k_3, k_1] + \tfrac{1}{3}[k_3, k_2, k_1] \\
&\quad + \tfrac{1}{12}[k_2, k_3, k_2, k_1] + \tfrac{1}{24}[k_3, k_1, k_3, k_1] - \tfrac{1}{12}[k_3, k_2, k_2, k_1] \\
j_3 &= k_3 - \tfrac{1}{2}[k_3, k_2] - \tfrac{1}{2}[k_4, k_1] - \tfrac{1}{12}[k_2, k_3, k_2] - \tfrac{1}{6}[k_2, k_4, k_1] \\
&\quad + \tfrac{1}{12}[k_3, k_3, k_1] + \tfrac{1}{3}[k_4, k_2, k_1] + \tfrac{1}{12}[k_3, k_2, k_3, k_1] - \tfrac{1}{12}[k_3, k_3, k_2, k_1] \\
j_4 &= k_4 - \tfrac{1}{2}[k_4, k_2] - \tfrac{1}{2}[k_5, k_1] - \tfrac{1}{12}[k_2, k_4, k_2] + \tfrac{1}{12}[k_3, k_3, k_2] \\
&\quad - \tfrac{1}{6}[k_3, k_4, k_1] + \tfrac{1}{3}[k_4, k_3, k_1] + \tfrac{1}{24}[k_3, k_2, k_3, k_2] \\
j_5 &= k_5 - \tfrac{1}{2}[k_4, k_3] - \tfrac{1}{2}[k_5, k_2] - \tfrac{1}{6}[k_3, k_4, k_2] + \tfrac{1}{3}[k_4, k_3, k_2] \\
j_6 &= k_6 - \tfrac{1}{2}[k_5, k_3] - \tfrac{1}{12}[k_3, k_4, k_3].
\end{aligned} \tag{C.1}$$

When inverted, we have the results for k in terms of j, through total rank six:

$$\begin{aligned}
k_1 &= j_1 + \tfrac{1}{2}[j_2, j_1] - \tfrac{1}{6}[j_1, j_3, j_1] + \tfrac{1}{6}[j_2, j_2, j_1] \\
&\quad - \tfrac{1}{8}[j_1, j_3, j_2, j_1] - \tfrac{1}{24}[j_2, j_1, j_3, j_1] + \tfrac{1}{24}[j_2, j_2, j_2, j_1] \\
k_2 &= j_2 + \tfrac{1}{2}[j_3, j_1] - \tfrac{1}{6}[j_1, j_4, j_1] - \tfrac{1}{12}[j_2, j_3, j_1] + \tfrac{1}{6}[j_3, j_2, j_1] \\
&\quad - \tfrac{1}{24}[j_2, j_3, j_2, j_1] - \tfrac{1}{24}[j_3, j_1, j_3, j_1] + \tfrac{1}{24}[j_3, j_2, j_2, j_1] \\
k_3 &= j_3 + \tfrac{1}{2}[j_3, j_2] + \tfrac{1}{2}[j_4, j_1] - \tfrac{1}{6}[j_2, j_3, j_2] - \tfrac{1}{3}[j_2, j_4, j_1] \\
&\quad + \tfrac{1}{6}[j_3, j_3, j_1] + \tfrac{1}{6}[j_4, j_2, j_1] + \tfrac{1}{24}[j_2, j_2, j_3, j_2] \\
&\quad - \tfrac{1}{8}[j_2, j_3, j_3, j_1] + \tfrac{1}{24}[j_3, j_2, j_3, j_1] + \tfrac{1}{24}[j_3, j_3, j_2, j_1] \\
k_4 &= j_4 + \tfrac{1}{2}[j_4, j_2] + \tfrac{1}{2}[j_5, j_1] - \tfrac{1}{6}[j_2, j_4, j_2] - \tfrac{1}{12}[j_3, j_3, j_2] + \tfrac{1}{6}[j_3, j_4, j_1] \\
&\quad - \tfrac{1}{12}[j_4, j_3, j_1] + \tfrac{1}{24}[j_3, j_2, j_3, j_2] - \tfrac{1}{24}[j_3, j_3, j_3, j_1] \\
k_5 &= j_5 + \tfrac{1}{2}[j_5, j_2] + \tfrac{1}{2}[j_4, j_3] - \tfrac{1}{12}[j_3, j_4, j_2] \\
&\quad - \tfrac{1}{12}[j_4, j_3, j_2] + \tfrac{1}{24}[j_3, j_3, j_3, j_2] \\
k_6 &= j_6 + \tfrac{1}{2}[j_5, j_3] + \tfrac{1}{12}[j_3, j_4, j_3].
\end{aligned} \tag{C.2}$$

The results in the tables of Appendix 4.B, as well as the examples of Sections 4.4.2 and 4.4.4, may be checked against this by setting the appropriate j terms to zero and discarding the Poisson brackets whose total rank exceeds six.

4.6 Acknowledgements

This paper is based on the dissertation [10] of one of the authors. We would also like to thank the Inference Corporation, who provided much assistance in understanding and applying *SMP*. This work was supported by U.S. Department of Energy contract number DE-AS05-80ER-10666.

4.7 References

[1] N. Bourbaki, *Elements of Mathematics, Lie Groups and Lie Algebras* (Addison-Wesley, Reading, Massachusetts, 1971); Part I: Chapters 1–3.

[2] D. R. Douglas, Ph.D. dissertation, University of Maryland Physics Department Technical Report (1982).

[3] A.J. Dragt, Lectures on Nonlinear Orbit Dynamics, in: *Physics of High Energy Particle Accelerators*, Proceedings of the 1981 Summer School on High Energy Particle Accelerators, New York, R.A. Carrigan, F.R. Huson, and M. Month, eds., American Institute of Physics, Conference Proceedings Number 87, 1982, p. 147.

[4] A.J. Dragt, and J.M. Finn, Lie series and invariant functions for analytic symplectic maps, *J. Math. Phys.* **17**, 2215 (1976).

[5] A.J. Dragt, and E. Forest, Computation of nonlinear behavior of Hamiltonian systems using Lie algebraic methods, *J. Math. Phys.* **24**, 2734 (1983).

[6] A.J. Dragt *et al.*, MARYLIE 3.0 manual, University of Maryland Technical Report (1988).

[7] A.J. Dragt, L.M. Healy, F.Neri, R.D. Ryne, D.R. Douglas, and E.Forest, MARYLIE 3.0 —A Program for Nonlinear Analysis of Accelerators and Beamline Lattices, *IEEE Trans. Nucl. Sci.* **NS-32**, 2311 (1985).

[8] A.J. Dragt, F. Neri, G. Rangarajan, D. Douglas, L.M. Healy, and R.D. Ryne, Lie algebraic treatment of linear and nonlinear beam dynamics, *Ann. Rev. Nucl. Part. Sci.* **38**, 455–496 (1988).

[9] M. Hausner and J. Schwartz, *Lie Groups Lie Algebras* (Gordon and Breach, New York, 1968).

[10] L.M. Healy, Ph.D. dissertation, University of Maryland Physics Department Technical Report (1986).

[11] L.M. Healy and A.J. Dragt, Lie algebraic methods for treating lattice parameter errors in accelerators. In: Proceedings of the 1987 IEEE Particle Accelerator Conference Vol. 2, (1987), p. 1060 *et seq.*

[12] L.M. Healy and A.J. Dragt, Computation of nonlinear behavior of inhomogeneous Hamiltonian systems using Lie algebraic methods (in preparation).

[13] *SMP Manual*, Inference Corporation, 1985.

[14] F.J. Murray and K.S. Miller, *Existence Theorems* (New York University Press, New York, 1954).

[15] V.S. Varadarajan, *Lie Groups, Lie Algebras, and their Representations* (Springer-Verlag, New York, 1984).

5

Dispersion-diffraction coupling in anisotropic media and ambiguity function generation

Moshe Nazarathy, Joseph W. Goodman, and Mark Kauderer

ABSTRACT We show that in the process of propagation of a space-time wavefield, a cross-ambiguity function of the spatial and temporal parts of the input distribution is generated, provided the $\omega - k$ *vector* dispersion relations of the medium exhibit a certain coupling between temporal dispersion and spatial anisotropy. This conclusion is obtained in the process of a comprehensive refinement of the canonical operator formalism and its application to the combined effects of diffraction and dispersion. This result opens some possibilities for novel signal processing techniques in the domain of ultrafast optics.

5.1 Introduction

In reference [1] we studied paraxial diffraction of monochromatic beams in birefringent media, generalizing the Huygens-Fresnel diffraction transforms to homogeneous anisotropic media. Our objective here is to extend that study to nonmonochromatic narrowband fields. We present a unified treatment in the space-time domain for pulsed beam propagation in media that are both anisotropic and dispersive. We point out the formal similarity between the mathematical manipulations of the spatial and temporal parts of the wavefield and indicate a new effect that involves a mixture of the two distinct diffraction and dispersion effects: the intrinsic generation of an ambiguity function between the spatial and temporal components of the field.

While the formal parallelism between dispersion and diffraction has been long known, the analogy has usually been limited to its first-order aspects. Thus, the phenomenon of beam walkoff in anisotropic media has been interpreted as an effective dispersion phenomenon in the spatial domain, due to discrepancy between effective spatial group and phase velocities. With the advent of ultrafast optical technology, understanding and combating the second-order phenomenon of temporal pulse distortion and spreading has become an ever more important task. Here we provide the operational analytical tools for treating dispersion as an effective diffraction mechanism in the temporal domain by means of Huygens-Fresnel Transforms, extending the canonical operator representation [2], [3] to the time-domain. The merit of this particular portion of the paper is mainly representational, applying the canonical operator formalism to the treatment of dispersion on the same footing as its initial use to the description of diffraction [1], facilitating the mathematical manipulations and allowing physical insight into the analogy between space and time quantities.

In ultrafast optics, spatial propagation in tilted configurations as induced by gratings or prisms is one of the mechanisms applied to combat pulse distortion [4], [5]. No unified theory exists at the moment addressing the common features of these effects and techniques. While not directly pursuing such a theory, this paper treats the related problem of pulse propagation in homogeneous anisotropic and dispersive media pointing out the yet unnoticed and experimentally undetected effects exhibited through the interaction between the spatial and temporal characteristics of the propagation process.

The main novel contribution of the paper consists in demonstrating that *in the process of propagation of a space-time wavefield, a cross-ambiguity function of the spatial and temporal parts of the input distribution is generated, provided the ω-k vector dispersion relations of the medium exhibit a certain coupling between temporal dispersion and spatial anisotropy.* This intriguing result opens some new possibilities for novel signal processing methods in the domain of ultrafast optics.

Essentially, the quality required of materials capable of generating the ambiguity function consists of the following: axial dispersion of temporal wavepackets varies with direction of propagation within the crystal, *i.e.* the group velocity is anisotropic; thus such materials exhibit ***anisotropic dispersion***. Alternatively, the material should exhibit ***dispersive anisotropy***, in the sense that its spatial anisotropy characteristics change with temporal frequency, *i.e.* the Poynting vector walkoff angle varies upon change of temporal frequency.

5.2 Dispersion relations in linear homogeneous media

We consider the most general type of linear homogeneous nonmagnetic medium. The constitutive relation for magnetic field quantities is simply given by

$$\boldsymbol{B} = \mu_0 \boldsymbol{H}. \tag{2.1}$$

The linear relationship between **D** and **E** is necessarily shift-invariant since the medium is homogeneous; this is most generally expressed as a tensorial convolution:

$$\mathbf{D}(\mathbf{r},t) = \int_{-\infty}^{\infty}\int_{-\infty}^{t} \epsilon_0\,\boldsymbol{\epsilon}(\mathbf{r}-\mathbf{r}',t-t')\cdot\mathbf{E}(\mathbf{r}',t')d\mathbf{r}'dt'. \tag{2.2}$$

Let us introduce 4-D Fourier transforms, *e.g.*,

$$\mathbf{E}(\mathbf{p},w) = \int_{-\infty}^{\infty} d\mathbf{r}\,dt\, e^{ik_0(\mathbf{p}\cdot\mathbf{r}-wt)}\mathbf{E}(\mathbf{r},t), \tag{2.3}$$

with fundamental spatio-temporal harmonics of the form $e^{ik_0(\mathbf{p}\cdot\mathbf{r}-wt)}$ where $\mathbf{p}$ and w are taken as normalized versions of the $\mathbf{k}$-vector and of the temporal angular frequency ω, respectively:

$$\mathbf{k} = k_0\mathbf{p}, \tag{2.4a}$$
$$\omega = k_0 w, \tag{2.4b}$$

where k_0 is some reference wavenumber, to be chosen later as the vacuum wavenumber corresponding to the midband frequency of the wavefield. In the frequency domain the space-time convolution (2.2) reduces to tensor multiplication:

$$\mathbf{D}(\mathbf{p},w) = \epsilon_0\boldsymbol{\epsilon}(\mathbf{p},w)\cdot\mathbf{E}(\mathbf{p},w). \tag{2.5}$$

It is apparent that the dielectric tensor might generally be dependent upon the wavevector in addition to its customary temporal frequency dependence.

An additional relation between $\mathbf{D}$ and $\mathbf{E}$ may be obtained from Maxwell's equations. Written in the Fourier domain, *i.e.* applied to fields with spatio-temporal dependence of the form $e^{ik_0(\mathbf{p}\cdot\mathbf{r}-w)}$, Maxwell's equations read

$$\mathbf{p}\times\mathbf{E} = w\mu_0\mathbf{H}, \tag{2.6a}$$
$$\mathbf{p}\times\mathbf{H} = -w\mathbf{D}. \tag{2.6b}$$

Combining these equations with the constitutive relations (2.1) and (2.5) yields a wave equation for the complex amplitude of the electric field

$$\left[\left(\frac{w}{c}\right)^2\boldsymbol{\epsilon}(\mathbf{p},w) + \mathbf{p}\mathbf{p}^{\mathsf{T}} - \mathbf{1}\right]\cdot\mathbf{E} = 0, \tag{2.7}$$

where $\mathbf{p}^{\mathsf{T}}$ is the transpose of the column vector $\mathbf{p}$. To obtain nontrivial solutions of this equation we require

$$\det\left[\left(\frac{w}{c}\right)^2\boldsymbol{\epsilon}(\mathbf{p},w) + \mathbf{p}\mathbf{p}^{\mathsf{T}} - \mathbf{1}\right] = 0, \tag{2.8}$$

which is a generalized version of Fresnel's equation [6] relating the allowed values of temporal and spatial frequencies of the elementary plane wave fields composing a general plane wave spectrum. Fresnel's equation defines a surface

$$F(p_x, p_y, p_z, w) = 0 \tag{2.9}$$

in 4-D spatio-temporal frequency space $(\mathbf{p}, w)$.

In this paper we exclusively restrict the treatment to fields that are 2-D in space, i.e. uniform along a certain direction, say the x axis. Such a general wavefield $u(t, y, z)$ may be synthesized from a plane wave spectrum with all $\mathbf{p}$-vectors lying in the y–z plane. Hence, setting

$$p_x = 0, \tag{2.10}$$

the 4-D surface (2.9) reduces to a 3-D surface in the frequency space (w, p_y, p_z), to be called the *dispersion surface:*

$$F(0, p_y, p_z, w) = 0. \tag{2.11}$$

A functional dependence

$$p_z = p_z(w, p_y) \tag{2.12}$$

may be extracted from the dispersion surface. This last equation is called the dispersion relation, generalizing the usual 1-D k–ω (or in our notation, p–w) dispersion relation by incorporating spatial features, explicitly indicating the interrelation between the p_y and p_z spatial frequency components, and its variation with temporal frequency.

In fact, sectioning the dispersion surface along, say, plane p_z, yields the customary k–ω dispersion curve for plane waves propagating along the y axis. On the other hand, sectioning the dispersion surface across a plane $w =$ constant yields the wavevector curve describing the anisotropy of the medium, corresponding to a 2-D section across the customary wavevector surface describing the monochromatic Fresnel equation.

For a nondispersive medium the first section is a straight line through the origin, while the second section is a circle centered on the w-axis. Thus, the dispersion surface for an isotropic nondisperive medium, say free-space, consists of cone with its vertex at the origin and its axis along the w-axis.

It is worthwhile visualizing the dispersion surfaces for dispersive-isotropic and nondispersive-anisotropic media. The former consists of a surface of revolution obtained by rotating the p–w dispersion relation around the w-axis. All $w =$ constant sections through this surface are circles, but the radii of these circles do not vary linearly with w but rather according to the p–w law. The latter consists of a conical surface where the intersections with planes orthogonal to the conical axis w consist of ellipses with the same orientation.

For more general anisotropic-dispersive cases that will be of interest here, the dispersion surfaces become more difficult to visualize. For example, considering an anisotropic material in which the orientation of the principal axes of the dielectric tensor varies with frequency, its dispersion surface would be obtained by applying a torsion around the w-axis to the last surface described above, such that the intersection ellipses are not parallel but rather, their axes gradually rotate as one proceeds along the w-axis. It is then expected that sections through planes passing through the ω-axis are not straight lines any more, *i.e.* such a material inherently exhibits temporal dispersion by virtue of the fact that its spatial anisotropy characteristics vary with frequency.

5.3 Plane wave spectrum representation

Let us take a double Fourier Transform (FT) of the transverse slices of the wavefield across z = constant sections. This amounts to performing two separate 1-D FT's over the variables t and y, while keeping z as a parameter (See the Appendix for the definition of the 1-D FT)

$$\begin{aligned} U(w,p_y,z) &= {}^{p_y}\mathbf{F}_y\ {}^{w}\mathbf{F}_t^{-1}u(t,y,z) \\ &= (i\lambda_0)^{-1}\int dy\,dt\,e^{-ik_0(p_y y-wt)}u(t,y,z). \end{aligned} \tag{3.1}$$

Notice the explicit notation of initial and final variables in the FT as right subscripts and left superscripts, respectively, as well as the fact that the kernel of the FT includes a factor k_0 in the exponent and a normalization factor in front of the integral. Also, the superscript -1 on the temporal FT corresponds to the negative sign associated with the phase $-wt$ in the kernel. The corresponding inverse double FT is then given by

$$\begin{aligned} u(t,y,z) &= {}^{y}\mathbf{F}_{p_y}^{-1}\ {}^{t}\mathbf{F}_w U(w,p_y,z) \\ &= (i\lambda_0)^{-1}\int dp_y\,dw\,e^{ik_0(p_y y-wt)}U(w,p_y,z). \end{aligned} \tag{3.2}$$

The plane wave spectrum representation of a general wavefield $u(t,y,z)$ is obtained by a superposition of plane wave components running over spatial and temporal frequencies

$$u(t,y,z) = (i\lambda_0)^{-1}\int dw\,dp_y\,e^{ik_0(yp_y+zp_z(w,p_y)-wt)}U(w,p_y,0). \tag{3.3}$$

Notice that the plane wave components in the superposition are labelled by their temporal frequency w, as well as by their spatial transverse frequency p_y. The weight factors in this superposition are given by the double FT of the $z=0$ distribution, $U(w,p_y,0)$, as may be readily verified by the substitution $z=0$. The axial spatial frequency p_z which determines the axial phaseshift $p_z z$ is expressed according to the dispersion relation (2.12).

Using the definitions of forward and backward FT, the plane wave spectrum representation yields

$$U(w,p_y,z) = e^{ik_0 z p_z(w,p_y)}U(w,p_y,0). \tag{3.4}$$

Thus, viewing the medium as a linear system in two variables (y, t) with the input and output over the planes $z = 0$ and $z =$ constant respectively, it follows that $e^{ik_0 z p_z(p_y, w)}$ is the transfer function in the spatio-temporal frequency domain, generalizing the concepts of spatial transfer function for monochromatic propagation in free-space [7], and in anisotropic homogeneous media [1].

In order to facilitate the description of beam-like pulsed wavefields let us factor out a plane wave dependence, introducing complex envelopes $\tilde{u}(t, y, z)$ (also called baseband signals), as defined by

$$u(t, y, z) = \tilde{u}(t, y, z)e^{ik_0(\bar{p}_y y + \bar{p}_z z - \bar{w}t)}, \tag{3.5}$$

where $(\bar{p}_y, \bar{p}_z, \bar{w})$ are midband values for the spatial and temporal frequency content of the wavefield. Taking now a transverse FT and denoting the resulting spectrum of the baseband signal by a capital letter with a tilde, we have

$$\begin{aligned} \tilde{U}(w, p_y, z) &= {}^{w}\mathbf{F}_t^{-1}\, {}^{p_y}\mathbf{F}_y \tilde{u}(t, y, z) \\ &= e^{ik_0 \bar{p}_z z} U(w + \bar{w}, p_y + \bar{p}_y, z). \end{aligned} \tag{3.6}$$

Reformulating the transfer function relation (3.4) in terms of the baseband signals yields

$$\tilde{U}(w, p_y, z) = \tilde{H}(w, p_y)\tilde{U}(w, p_y, 0), \tag{3.7}$$

where

$$\tilde{H}(w, p_y) = \exp(ik_0 z[p_z(w + \bar{w}, p_y + \bar{p}_y) - \bar{p}_z]), \tag{3.8}$$

is the *baseband transfer function.*

Rewriting Eq. (3.7) in terms of the actual spatio-temporal baseband signals we obtain

$$\tilde{u}(t, y, z) = {}^{t}\mathbf{F}_w\, {}^{y}\mathbf{F}_{p_y}^{-1} \tilde{H}(w, p_y)\, {}^{p_y}\mathbf{F}_y\, {}^{w}\mathbf{F}_t^{-1} \tilde{u}(t, y, 0), \tag{3.9}$$

or

$$\tilde{u}(t, y, z) = \tilde{\mathbf{T}} \tilde{u}(t, y, 0) \tag{3.10}$$

where $\tilde{\mathbf{T}}$ is the *Baseband Transfer Operator* (BTO),

$$\tilde{\mathbf{T}} = {}^{t}\mathbf{F}_w\, {}^{y}\mathbf{F}_{p_y}^{-1} \tilde{H}(w, p_y)\, {}^{p_y}\mathbf{F}_y\, {}^{w}\mathbf{F}_t^{-1} \tag{3.11}$$

directly yielding the space-time domain input-output transformation for the baseband signals.

We now proceed to apply the Slowly-Varying Envelope Approximation (SVEA), which assumes the signal is narrowband in the w–p_y frequency domain, with its spectrum concentrated around $(\bar{p}_y, \bar{w})$. Thus, we expand the exponent in the baseband transfer function (3.8) in a Taylor power series, explicitly retaining terms up to second-order

$$\tilde{H}(w, p_y) = \exp(ik_0 z[\bar{\alpha}_2 w - \tfrac{1}{2}\bar{\beta}_2 w^2 - \bar{\alpha}_1 p_y - \tfrac{1}{2}\bar{\beta}_1 p_y^2 + \bar{\gamma} p_y w]), \tag{3.12}$$

where the coefficients $\bar{\alpha}$, $\bar{\beta}$ and $\bar{\gamma}$ are given by partial derivatives of the dispersion relation (2.12)

$$\bar{\alpha}_1 = -\left.\frac{\partial p_z}{\partial p_y}\right|_{\substack{p_y=\bar{p}_y \\ w=\bar{w}}}, \qquad \bar{\beta}_1 = -\left.\frac{\partial^2 p_z}{\partial p_y^2}\right|_{\substack{p_y=\bar{p}_y \\ w=\bar{w}}}, \tag{3.13a}$$

$$\bar{\alpha}_2 = +\left.\frac{\partial p_z}{\partial w}\right|_{\substack{p_y=\bar{p}_y \\ w=\bar{w}}}, \qquad \bar{\beta}_2 = -\left.\frac{\partial^2 p_z}{\partial w^2}\right|_{\substack{p_y=\bar{p}_y \\ w=\bar{w}}}, \tag{3.13b}$$

$$\bar{\gamma} = \left.\frac{\partial^2 p_z}{\partial p_y \partial w}\right|_{\substack{p_y=\bar{p}_y \\ w=\bar{w}}}. \tag{3.13c}$$

At this point the physics of the problem has been incorporated in the formulation and we proceed with the mathematical simplification and interpretation of the results.

5.4 Canonical operator formulation

Since the expressions obtained in the last section contain linear and quadratic phase factors as well as Fourier transformations it is most convenient to use the canonical operator formalism [2], [1] that was specifically developed to simplify and systematize the manipulations of such expressions. Using the canonical operator definitions reviewed in the Appendix, the baseband transfer function may be cast in the form

$$\widetilde{H}(w, p_y) = \mathbf{G}_{p_y}[-z\bar{\alpha}_1]\mathbf{Q}_{p_y}[-z\bar{\beta}_1]e^{ik_0\bar{\gamma}p_y w}\mathbf{G}_w[z\bar{\alpha}_2]\mathbf{Q}_w[-z\bar{\beta}_2], \tag{4.1}$$

where $\mathbf{G}$ and $\mathbf{Q}$ denote linear and quadratic phase factors respectively. Substituting the last expression into Eq. (3.11) and commuting some of the time-dependent operators with some of the space-dependent operators yields the following operator string for the BTO

$$\begin{aligned} \widetilde{\boldsymbol{T}} = {} & {}^{y}\mathbf{F}_{p_y}^{-1}\mathbf{G}_{p_y}[-z\bar{\alpha}_1]\mathbf{Q}_{p_y}[-z\bar{\beta}_1]\ {}^{t}\mathbf{F}_w e^{ik_0 z\bar{\gamma}p_y w}\ {}^{p_y}\mathbf{F}_y \\ & \times\mathbf{G}_w[z\bar{\alpha}_2]\mathbf{Q}_w[-z\bar{\beta}_2]\ {}^{w}\mathbf{F}_t^{-1}. \end{aligned} \tag{4.2}$$

5.5 Uncoupling of anisotropy and dispersion

In this section we show that for media satisfying

$$\bar{\gamma} = \frac{\partial^2 p_z}{\partial p_y \partial w} = 0 \tag{5.1}$$

the phenomena of anisotropy and dispersion become uncoupled.

Mathematically, discarding the exponential middle term in Eq. (4.2) and commuting the middle space and time-dependent FT's, yields

$$\widetilde{T} = {}^{y}\mathbf{F}^{-1}_{p_y}\mathbf{G}_{p_y}[-z\bar{\alpha}_1]\mathbf{Q}_{p_y}[-z\bar{\beta}_1]\,{}^{p_y}\mathbf{F}_y\,{}^{t}\mathbf{F}_w\mathbf{G}_w[z\bar{\alpha}_2]\mathbf{Q}_w[-z\bar{\beta}_2]\,{}^{w}\mathbf{F}_t^{-1}. \tag{5.2}$$

Applying some of the operator algebra rules outlined in the Appendix, we further reduce the BTO to a combination of free space Huygens Transforms $\mathbf{R}$, and shift operators $\mathbf{S}$:

$$\widetilde{T} = \mathbf{S}_y[z\bar{\alpha}_1]\mathbf{R}_y[z\bar{\beta}_1]\mathbf{S}_t[z\bar{\alpha}_2]\mathbf{R}_t[z\bar{\beta}_2]. \tag{5.3a}$$

Thus, the BTO may be partitioned into two operators, a spatial BTO $\widetilde{T}_t^{(1)}$ and a temporal BTO $\widetilde{T}_t^{(2)}$:

$$\widetilde{T} = \widetilde{T}_y^{(1)}\widetilde{T}_t^{(2)}, \tag{5.3b}$$

where

$$\widetilde{T}_y^{(1)} = \mathbf{S}_y[z\bar{\alpha}_1]\mathbf{R}_y[z\bar{\beta}_1], \tag{5.4a}$$

$$\widetilde{T}_t^{(2)} = \mathbf{S}_t[z\bar{\alpha}_2]\mathbf{R}_t[z\bar{\beta}_2]. \tag{5.4b}$$

The Huygens Transforms $\mathbf{R}$ represent the effects of diffraction in the spatial domain and pulse spreading in the temporal domain, respectively, while the shift operators $\mathbf{S}$ describe a pulse delay according to a group velocity in a dispersive medium and the phenomenon of beam walkoff in the presence of anisotropy [1], respectively. For a separable input, *i.e.* an input given by a product of a temporal and a spatial part

$$\tilde{u}(t, y, 0) = u_2(t)u_1(y), \tag{5.5}$$

the output is also separable, with its spatial and temporal parts being obtained by applying the spatial and temporal BTO's on the respective parts of the input. Thus, the whole wavefield is separable,

$$\tilde{u}(t, y, z) = [\widetilde{T}_t^{(2)}u_2(t)][\widetilde{T}_y^{(1)}u_1(y)], \tag{5.6}$$

with the propagation of the temporal and spatial parts separately governed by the respective BTO's. In particular, consider the monochromatic case which is characterized by a time-independent complex envelope, *i.e.*

$$u_2(t) = 1. \tag{5.7}$$

In this case only spatial effects are present as described by a space-dependent operator:

$$\widetilde{T}u_1(y) = \widetilde{T}_y^{(1)}u_1(y) = u_1^{\mathrm{diff}}(y). \tag{5.8}$$

The dual case is obtained by eliminating the transverse spatial dependence, *i.e.* taking

$$u_1(y) = 1 \tag{5.9}$$

corresponding to 1-D propagation of plane wavefront pulses, mathematically described by a time-dependent operator

$$\widetilde{\boldsymbol{T}} u_2(t) = \widetilde{\boldsymbol{T}}_t^{(2)} u_2(t) = u_2^{\mathrm{disp}}(t). \tag{5.10}$$

These results form the rigorous basis of the spatial-temporal analogy, whereby temporal quantities pertaining to dispersion are fully analogous to spatial quantities pertaining to diffraction.

5.6 Diffraction-dispersion analogy

In this section we briefly explore the formal analogy between dispersion and diffraction, under the condition (5.1), as exhibited in the formal similarity of expressions (5.4*a*) and (5.4*b*) describing spatial diffraction of monochromatic fields and temporal dispersion.

As discussed in Ref. [1], spatial diffraction of monochromatic fields in anisotropic media is equivalent to a free-space propagation through a distance $\bar{\beta}_1 z$ yielding beam spreading as well as to lateral shift $\bar{\alpha}_1 z$, describing beam walkoff. Thus, $\bar{\alpha}_1$ is the slope of the walkoff angle, while $\bar{\beta}_1$ is a diffraction-compressing factor given by the quotient of the effective propagation length and the actual propagation distance z.

Considering time-domain evolution as described by the transfer operator (5.4*b*), the major effect of temporal dispersion is the introduction of a time delay given by the label of the shift operator,

$$\bar{\alpha}_2 z = \Delta\tau = \frac{\partial p_z}{\partial w} z \tag{6.1a}$$

corresponding to a group velocity *along the z axis*,

$$\nu_{gz} = \frac{z}{\Delta\tau} = \bar{\alpha}_2^{-1} = \left(\left. \frac{\partial p_z}{\partial w} \right|_{\substack{p_y=\bar{p}_y \\ w=\bar{w}}} \right)^{-1} = \left. \frac{\partial \omega}{\partial k_z} \right|_{k=\bar{k}_z}. \tag{6.1b}$$

It is instructive to show the consistency of this z-oriented group velocity expression with the assumption of tilted propagation: When evaluating the partial derivative in Eq. (6.1*b*) one fixes the slope at $p = \bar{p}_y$. This in effect singles out a monodirectional sub-wavepacket which propagates spatially as a tilted plane wave with its peak intensity plane wavefront sweeping along the tilted fixed axis of propagation at velocity $\partial\omega/\partial k$, while simultaneously sweeping along the z axis at a speed which is by a factor of $\cos^{-1}\theta$ larger than the corresponding distance along the tilted propagation direction, where $\cos\theta$ is the z-direction cosine of the normal to the plane wavepacket. Thus

$$\nu_{gz} = \frac{\nu_g}{\cos\theta} = \frac{\partial\omega}{\partial k\cos\theta} = \frac{\partial\omega}{\partial k_z}, \tag{6.1c}$$

verifying equation (6.1*b*).

The temporal Huygens transform with label $z\bar{\beta}_2$ in Eq. (5.4*b*) describes temporal pulse distortion effects (in particular compression and decompression of Gaussian pulses). This temporal "free-space propagation" transformation amounts to considering a transverse spatial distribution with a profile identical to the initial temporal pulse profile, and letting this spatial distribution diffract through an effective free-space system along a distance given by $z\bar{\beta}_2$. The meaning of this parameter may be seen by writing

$$\bar{\beta}_2 z = \frac{\partial}{\partial\omega}\left(\frac{\partial p_z}{\partial\omega}z\right) = \frac{\partial\Delta\tau}{\partial\omega} = \frac{\partial}{\partial\omega}\frac{z}{\nu_g}. \tag{6.1d}$$

Thus, this parameter is numerically equal to the relative time lag with which two narrowband pulses, spaced a unit frequency apart and emitted simultaneously at the input, arrive at the output plane.

We may now consider the significance of condition (5.1) which leads to noninteracting spatial and temporal features. This condition implies that $\nu_g^{-1} = \partial p_z/\partial w|_{\bar{w}}$ is independent of $\bar{p}_y$ *i.e.* the group velocity in the z-direction for temporal pulses propagating in a small range of tilted directions is roughly independent of this tilt. However, nothing precludes the phase velocity to be dependent upon direction, *i.e* the medium could be assumed anisotropic while temporal dispersion exhibits isotropy.

Alternatively, condition (5.1) means that $\partial p_z/\partial y_y|_{\bar{p}_y}$, which is the Poynting vector walkoff slope, is independent of ω; *i.e.* the beam walkoff slope does not vary with temporal frequency, implying that the spatial anisotropy characteristics are nondispersive.

Thus, media which exhibit uncoupling of the space and time features can be characterized in two dual ways either as *isotropically-dispersive* or as *non-dispersively anisotropic*.

In the next section we proceed to derive the interesting effects occurring for $\bar{\gamma} \neq 0$, *i.e.* in materials which are *anisotropically-dispersive* or alternatively *dispersively-anisotropic*.

5.7 Diffraction-dispersion interaction

In this section we show how in the presence of the space-time coupling factor $e^{ik_0 z\bar{\gamma}p_y w}$, the BTO no longer decouples into noninteracting spatial and temporal parts. Rather, the temporal and spatial features interact to produce a convolutional cross-ambiguity function of the temporal signal with the spectrum of the spatial signal.

Inserting the identity operators ${}^{p_y}\mathbf{F}_y\ {}^{y}\mathbf{F}_{p_y}^{-1}$ and ${}^{w}\mathbf{F}_t\ {}^{t}\mathbf{F}_w$ into the operator chain and omitting for simplicity the explicit notation of operator brackets and parameters we have

$$\widetilde{T} = {}^{y}\mathbf{F}_{p_y}^{-1}\mathbf{G}_{p_y}\mathbf{Q}_{p_y}\ {}^{p_y}\mathbf{F}_y\ {}^{y}\mathbf{F}_{p_y}^{-1}\ {}^{t}\mathbf{F}_w e^{ik_0 z\bar{\gamma}p_y w}\ {}^{p_y}\mathbf{F}_y\ {}^{w}\mathbf{F}_t^{-1}\ {}^{t}\mathbf{F}_w\mathbf{G}_w\mathbf{Q}_w\ {}^{w}\mathbf{F}_t^{-1} \tag{7.1}$$

or

$$\widetilde{T} = \widetilde{T}_y^{(1)}\mathbf{A}\widetilde{T}_t^{(2)}, \tag{7.2}$$

where

$$\begin{aligned} \mathbf{A} &= {}^{y}\mathbf{F}_{p_y}^{-1}\ {}^{t}\mathbf{F}_w e^{ik_0 z\bar{\gamma}p_y w}\ {}^{p_y}\mathbf{F}_y\ {}^{w}\mathbf{F}_t^{-1} & (7.3)\\ &= {}^{y}\mathbf{F}_{p_y}^{-1}\mathbf{S}_t[\bar{\gamma}zp_y]\ {}^{t}\mathbf{F}_w\ {}^{p_y}\mathbf{F}_y\ {}^{w}\mathbf{F}_t^{-1} & \\ &= {}^{y}\mathbf{F}_{p_y}^{-1}\mathbf{S}_t[\bar{\gamma}zp_y]\ {}^{p_y}\mathbf{F}_y. & (7.4) \end{aligned}$$

The meaning of the last form for $\mathbf{A}$ is as follows: Let $\psi(t,y)$ be a general space-time function, and let

$$\Phi(t,p_y) = {}^{p_y}\mathbf{F}_y\psi(t,y) \tag{7.5}$$

be its spatial FT. Applying the $\mathbf{A}$ operator yields:

$$\mathbf{A}\psi(t,y) = {}^{y}\mathbf{F}_{p_y}^{-1}\Phi(t-\bar{\gamma}zp_y,p_y).$$

No further simplification is possible here, but we now show that a convolutional-ambiguity function arises in the case of separable fields.

Consider a separable function in the variables y and t:

$$\psi(t,y) = \psi_1(y)\psi_2(t), \tag{7.6}$$

and denote 1-D FT's by corresponding uppercase letters:

$$\Psi_1(p_y) = {}^{p_y}\mathbf{F}_y\psi_1(y) \tag{7.7a}$$

and

$$\Psi_2(w) = {}^{w}\mathbf{F}_t^{-1}\psi_2(t). \tag{7.7b}$$

Applying the $\mathbf{A}$ operator on $\psi(t,y)$ yields

$$\begin{aligned} \mathbf{A}\psi_1(y)\psi_2(t) &= {}^{y}\mathbf{F}_{p_y}^{-1}\mathbf{S}_t[\bar{\gamma}zp_y]\ {}^{p_y}\mathbf{F}_y\psi_1(y)\psi_2(t) & (7.8)\\ &= {}^{y}\mathbf{F}_{p_y}^{-1}\{\mathbf{S}_t[\bar{\gamma}zp_y]\psi_2(t)\}\ {}^{p_y}\mathbf{F}_y\psi_1(y) & \\ &= {}^{y}\mathbf{F}_{p_y}^{-1}\psi_2(t-\bar{\gamma}zp_y)\Psi_1(p_y) & \\ &= (\bar{\gamma}z)^{-1/2}\ {}^{y}\mathbf{F}_{p_y}^{-1}\mathbf{V}_{p_y}[\bar{\gamma}z]\psi_2(t-p_y)\Psi_1(p_y/\bar{\gamma}z) & \\ &= \mathbf{V}_y[1/\bar{\gamma}z]\ {}^{y}\mathbf{F}_{p_y}^{-1}\psi_2(t-p_y)\Psi_1^{\mathrm{scal}}(p_y), & \end{aligned}$$

where we introduced the definition

$$\Psi_1^{\mathrm{scal}}(p_y) = (\bar{\gamma}z)^{-1/2}\Psi_1(p_y/\bar{\gamma}z) = {}^{p_y}\mathbf{F}_y\mathbf{V}[\bar{\gamma}z]\psi_1(y). \tag{7.9}$$

To understand the meaning of this result let us introduce the following definition for the *Convolutional-Ambiguity Function* (CAF) of two given functions f and g:

$$\begin{aligned} a(\nu,\mu) &= (i\lambda_0)^{-1/2}\int dq\, e^{ik_0\nu q} f(\mu-q)g(q) \\ &= {}^{\nu}\mathbf{F}_q^{-1} f(\mu-q)g(q). \end{aligned} \tag{7.10}$$

This representation was named *Doppler-frequency selective convolution* in Ref. [8], as it reduces along the μ-axis of the ν–μ plane to a convolution of f and g whereas along the ν-axis it yields the Doppler difference frequency of the two signals, in case the signals are narrowband. Our definition differs only superficially from that of Ref. [8] through the use of an "optical" kernel $e^{ik_0\nu q}$ instead of $e^{-j2\pi\nu q}$, amounting to a scaling of the ν coordinate by a factor $-\lambda_0$. It is convenient to introduce the following compact notation for the CAF of two given functions f and g,

$$a(\nu,\mu) = CAF_{\nu,\mu}\{f;g\} \tag{7.11}$$

indicating the two variables ν and μ in subscript form. The notation implies that the final result does not depend upon the dummy variables in which the functions f and g are initially displayed. With these definitions and notations, it is apparent that the effect of the $\mathbf{A}$ operator is the generation of a scaled CAF of one function with the scaled spectrum of the other function

$$\mathbf{A}\psi_1(y)\psi_2(t) = \mathbf{V}_y[1/\bar{\gamma}z]CAF_{y,t}\{\psi_2;\psi_1^{\mathrm{scal}}\}. \tag{7.12}$$

We are now in a position to derive the overall transformation of the separable field through the BTO: Let the input baseband signal be separable as described by Eq. (5.5); then the output baseband signal is given by

$$\begin{aligned} \tilde{u}(t,y,z) &= \tilde{\boldsymbol{T}}\tilde{u}(t,y.0) \\ &= \tilde{\boldsymbol{T}}_y^{(1)}\mathbf{A}\tilde{\boldsymbol{T}}_t^{(2)}u_2(t)u_1(y) \\ &= \tilde{\boldsymbol{T}}_y^{(1)}\mathbf{A}u_2^{\mathrm{disp}}(t)u_1(y) \\ &= \tilde{\boldsymbol{T}}_y^{(1)}\mathbf{V}_y[1/\bar{\gamma}z]CAF_{y,t}\left\{u_2^{\mathrm{disp}};U_1^{\mathrm{scal}}\right\} \\ &= \tilde{\boldsymbol{T}}_y^{(1)}\mathbf{V}_y[1/\bar{\gamma}z]CAF_{y,t}\left\{\tilde{\boldsymbol{T}}_t^{(2)}u_2(t);\mathbf{F}_y\mathbf{V}_y[z\bar{\gamma}]u_1(y)\right\}, \end{aligned} \tag{7.13}$$

i.e., we obtain a diffracted and spatially-scaled convolutional ambiguity function of the dispersed version of the time part with the scaled spectrum of the spatial part. This overall transformation is our main result in this paper.

5.8 Space-time duality

The duality between space and time was already revealed in the analogous forms for the diffraction and dispersion operators $\widetilde{T}_y^{(1)}$ and $\widetilde{T}_t^{(2)}$.

In fact this duality is further preserved in the case when $\bar{\gamma} \neq 0$ *i.e.* when diffraction and dispersion become coupled, as indicated by the symmetric form under which p_y and w appear in the mixed factor $e^{ik_0 z\bar{\gamma}p_y w}$. Formally all our equations are invariant under the following symmetry substitutions

$$\begin{aligned} t &\leftrightarrow y \\ w &\leftrightarrow -x \\ \text{index } 1 &\leftrightarrow \text{index } 2 \\ {}^w\mathbf{F}_t^{-1} &\leftrightarrow {}^{p_y}\mathbf{F}_y. \end{aligned} \tag{8.1}$$

We start by redoing the derivation of the last section, regrouping the space and time dependent operators in opposing directions as compared with that derivation. Thus, the baseband transfer function (4.1) is written

$$\widetilde{H}(w, p_y) = \mathbf{G}_w[z\bar{\alpha}_2]\mathbf{Q}_w[-z\bar{\beta}_2]e^{ik_0 z\bar{\gamma}p_y w}\mathbf{G}_{p_y}[-z\bar{\alpha}_1]\mathbf{Q}_{p_y}[-z\bar{\beta}_1], \tag{8.2}$$

and the BTO becomes

$$\begin{aligned} \widetilde{T} &= {}^t\mathbf{F}_w \mathbf{G}_w[z\bar{\alpha}_2]\mathbf{Q}_w[-z\bar{\beta}_2]\ {}^y\mathbf{F}_{p_y}^{-1} e^{ik_0 zmp_y w}\ {}^w\mathbf{F}_t^{-1} \\ &\quad \times \mathbf{G}_{p_y}[-z\bar{\alpha}_1]\mathbf{Q}_{p_y}[-z\bar{\beta}_1]\ {}^{p_y}\mathbf{F}_y. \end{aligned} \tag{8.3}$$

The operator chain may further be manipulated into the form

$$\widetilde{T} = {}^t\mathbf{F}_w \mathbf{G}_w \mathbf{Q}_m\ {}^w\mathbf{F}_t^{-1}\ {}^t\mathbf{F}_w\ {}^y\mathbf{F}_{p_y}^{-1} e^{ik_0 z\bar{\gamma}p_y w}\ {}^w\mathbf{F}_t^{-1}\ {}^{p_y}\mathbf{F}_y\ {}^y\mathbf{F}_{p_y}^{-1}\mathbf{G}_{p_y}\mathbf{Q}_{p_y}\ {}^{p_y}\mathbf{F}_y, \tag{8.4}$$

or

$$\widetilde{T} = \widetilde{T}^{(2)}\mathbf{A}'\widetilde{T}^{(1)}, \tag{8.5}$$

where

$$\begin{aligned} \mathbf{A}' &= {}^t\mathbf{F}_w\ {}^y\mathbf{F}_{p_y}^{-1} e^{ik_0\bar{\gamma}p_y w}\ {}^w\mathbf{F}_t^{-1}\ {}^{p_y}\mathbf{F}_y && (8.6) \\ &= {}^t\mathbf{F}_w \mathbf{S}_y[-\bar{\gamma}zw]\ {}^y\mathbf{F}_{p_y}^{-1}\ {}^w\mathbf{F}_t^{-1}\ {}^{p_y}\mathbf{F}_y \\ &= {}^t\mathbf{F}_w \mathbf{S}_y[-\bar{\gamma}zw]\ {}^w\mathbf{F}_t^{-1}. && (8.7) \end{aligned}$$

We now show that the operator $\mathbf{A}'$ yields a correlational version of the ambiguity function. Applying the $\mathbf{A}'$ operator on a separable function $\psi(t, y)$ yields:

$$\begin{aligned} \mathbf{A}'\psi_1(y)\psi_2(t) &= {}^t\mathbf{F}_w \mathbf{S}_y[-\bar{\gamma}zw]\ {}^w\mathbf{F}_t^{-1}\psi_1(y)\psi_2(t) \\ &= {}^t\mathbf{F}_w\{\mathbf{S}_y[-\bar{\gamma}zw]\psi_1(y)\}\ {}^w\mathbf{F}_t^{-1}\psi_2(t) \\ &= {}^t\mathbf{F}_w\psi_1(y + \bar{\gamma}zw)\Psi_2(w) \\ &= (\bar{\gamma}z)^{-1/2}\ {}^t\mathbf{F}_w \mathbf{V}_w[\bar{\gamma}z]\psi_1(y + w)\Psi_2(w/\bar{\gamma}z) \\ &= \mathbf{V}_t[1/\bar{\gamma}z]\ {}^t\mathbf{F}_w\psi_1(y + w)\Psi_2^{\text{scal}}(w), \end{aligned}$$

where we defined

$$\Psi_2^{\mathrm{scal}}(w) = (\bar{\gamma}z)^{-1/2}\Psi_2(w/\bar{\gamma}z). \tag{8.8}$$

This result may be interpreted as a Correlational-Real Ambiguity Function (CRAF), a concept defined in Ref. [8] under the name *mean-frequency selective correlation*. Our definition of the CRAF of two given functions f and g is

$$CRAF_{\nu,\mu}\{f;g\} = (i\lambda_0)^{-1/2}\int dq\, e^{-ik_0\nu q} f(\mu+q)g(q) = {}^{\nu}\mathbf{F}_q f(\mu+q)g(q) \tag{8.9}$$

The effect of $\mathbf{A}'$ is therefore CRAF generation:

$$\mathbf{A}'\psi_1(y)\psi_2(t) = \mathbf{V}_t[1/\bar{\gamma}z]CRAF_{t,y}\{\psi_1;\psi_2^{\mathrm{scal}}\}. \tag{8.10}$$

The evolution of a separable baseband field is then described by:

$$\begin{aligned} u(t,y,z) &= \widetilde{T}u(t,y,0) \qquad (8.11)\\ &= \widetilde{\mathbf{T}}_t^{(2)}\mathbf{A}'\widetilde{\mathbf{T}}_y^{(1)}u_1(y)u_2(t)\\ &= \widetilde{\mathbf{T}}_t^{(2)}\mathbf{V}_t[1/\bar{\gamma}z]CRAF_{t,y}\{u_1^{\mathrm{diff}};U_2^{\mathrm{scal}}\}\\ &= \widetilde{\mathbf{T}}_t^{(2)}\mathbf{V}_t[1/\bar{\gamma}z]CRAF_{t,y}\left\{\widetilde{\mathbf{T}}_y^{(1)}u_1;\ {}^{w}\mathbf{F}_t^{-1}\mathbf{V}_t[\bar{\gamma}z]u_2\right\}, \end{aligned}$$

i.e., we obtain a dispersed and temporally-scaled ambiguity function of the diffracted version of the time part with the scaled spectrum of the spatial part.

5.9 Discussion

It has been shown that the joint action of dispersion and diffraction in certain materials results in the generation of the convolutional ambiguity function of certain transforms of the temporal and spatial parts of a separable input field.

A preliminary discussion of the structure of a processor which makes use of this effect to optically implement the ambiguity function of two given signals is given in Ref. [9]. To pursue this idea to the practical implementation of an optical processor much work remains to be done in actually identifying suitable crystals with dispersive-anisotropy (or anisotropic-dispersion) compatible with the requisite time resolution. The significance of the current results for the domain of ultrafast optics remains to be evaluated. The proposed processor and some other similar versions could constitute either a competing or a complementing technology to the recently considered class of acousto-optic processors [8] which implement the ambiguity function and other similar signal representations.

A particularly attractive prospect consists of converting the convolutional-ambiguity function processor into a temporal time-invariant filter which would compute convolutions with an impulse response that is spatially preset by means of a spatial-modulator.

5.A Appendix: Canonical operator algebra

5.A.1 Basic operator definitions [1], [2], [3]

Quadratic Phase	$\mathbf{Q}[c]u(x)$	$=$	$\exp(ik_0 cx^2/2)\, u(x)$
Scaling	$\mathbf{V}[a]u(x)$	$=$	$a^{1/2}u(ax)$
Fourier	$\mathbf{F}u(x)$	$=$	$(i\lambda_0)^{-1/2}\int dx \exp(ik_0 px)\, u(x)$
Huygens Transform (free-space propagation)	$\mathbf{R}[d]u(x_1)$	$=$	$(i\lambda_0 d)^{-1/2}\int dx_1 \times \exp(ik_0(x_2-x_1)^2/2d)\, u(x_1)$
Shift	$\mathbf{S}[m]u(x)$	$=$	$u(x-m)$
Linear Phase	$\mathbf{G}[p]u(x)$	$=$	$\exp(ik_0 px)\, u(x)$

5.A.2 Notation of variables

We denote the initial and final variables in an operator relation as right subscripts and left superscripts, respectively:

$$u_2(r) = {}^r\mathbf{T}_q u_1(q).$$

Thus, the Fourier operator as defined above is ${}^p\mathbf{F}_x$.

Sometimes we may drop one of the indices, writing for example $\mathbf{V}_x$.

Usually, we drop both indices when manipulating operator combinations, as in the Table, but in this paper it is necessary to keep track of indices to distinguish between spatial and temporal variables.

For the linear and quadratic phase factors, writing $\mathbf{Q}_x[c]$, $\mathbf{G}_x[p]$, and letting these symbols stand on their own rather than applying them to some functions, means we use the notation in a functional rather than an operational sense, *e.g.* $\mathbf{Q}_y(c)\mathbf{G}_x[p]$ denotes here the function

$$\mathbf{Q}_y[c]\mathbf{G}_x[p] = \exp ik_0(px + \tfrac{1}{2}cy^2).$$

5.A.3 Operator algebra

The relations satisfied by the operators in this Appendix are presented in the Table.

	V	**F**	**Q**	**R**	**G**	**S**
V	$\mathbf{V}[a_2]\mathbf{V}[a_1] = \mathbf{V}[a_2 a_1]$	$\mathbf{V}[a]\mathbf{F} = \mathbf{F}\mathbf{V}[\frac{1}{a}]$	$\mathbf{V}[a]\mathbf{Q}[c] = \mathbf{Q}[a^2 c]\mathbf{V}[a]$	$\mathbf{V}[a]\mathbf{R}[d] = \mathbf{R}[\frac{d}{a^2}]\mathbf{V}[a]$	$\mathbf{V}[a]\mathbf{G}[p] = \mathbf{G}[pa]\mathbf{V}[a]$	$\mathbf{V}[a]\mathbf{S}[m] = \mathbf{S}[ma^{-1}]\mathbf{V}[a]$
F	$\mathbf{F}\mathbf{V}[a] = \mathbf{V}[\frac{1}{a}]\mathbf{F}$	$\mathbf{F}\mathbf{F} = \mathbf{V}[-1]$	$\mathbf{F}\mathbf{Q}[c] = \mathbf{R}[-c]\mathbf{F}$	$\mathbf{F}\mathbf{R}[d] = \mathbf{Q}[-d]\mathbf{F}$	$\mathbf{F}\mathbf{G}[p] = \mathbf{S}[p]\mathbf{F}$	$\mathbf{F}\mathbf{S}[m] = \mathbf{G}[-m]\mathbf{F}$
Q	$\mathbf{Q}[c]\mathbf{V}[a] = \mathbf{V}[a]\mathbf{Q}[\frac{c}{a^2}]$	$\mathbf{Q}[c]\mathbf{F} = \mathbf{F}\mathbf{R}[-c]$	$\mathbf{Q}[c_2]\mathbf{Q}[c_1] = \mathbf{Q}[c_2 + c_1]$	$\mathbf{Q}[c]\mathbf{R}[d] = \mathbf{R}[(d^{-1} + c)^{-1}]$ $\times \mathbf{V}[1 + cd]\mathbf{Q}[(c^{-1} + d)^{-1}]$	$\mathbf{Q}[c]\mathbf{G}[p] = \mathbf{G}[p]\mathbf{Q}[c]$	$\mathbf{Q}[c]\mathbf{S}[m] = (\mathbf{Q}_m[-c])\mathbf{Q}[c]\mathbf{G}[-m]$
R	$\mathbf{R}[d]\mathbf{V}[a] = \mathbf{V}[a]\mathbf{R}[a^2 d]$	$\mathbf{R}[d]\mathbf{F} = \mathbf{F}\mathbf{Q}[-d]$	$\mathbf{R}[d]\mathbf{Q}[c] = \mathbf{Q}[(c^{-1} + d)^{-1}]$ $\times \mathbf{V}[(1 + cd)^{-1}]\mathbf{R}[(d^{-1} + c)^{-1}]$	$\mathbf{R}[d_2]\mathbf{R}[d_1] = \mathbf{R}[d_2 + d_1]$	$\mathbf{R}[d]\mathbf{G}[p] = (\mathbf{Q}_p[-d])$ $\times \mathbf{G}[p]\mathbf{S}[pd]\mathbf{R}[d]$	$\mathbf{R}[d]\mathbf{S}[m] = \mathbf{S}[m]\mathbf{R}[d]$
G	$\mathbf{G}[p]\mathbf{V}[a] = \mathbf{V}[a]\mathbf{G}[pa^{-1}]$	$\mathbf{G}[p]\mathbf{F} = \mathbf{F}\mathbf{S}[-p]$	$\mathbf{G}[p]\mathbf{Q}[c] = \mathbf{Q}[c]\mathbf{G}[p]$	$\mathbf{G}[p]\mathbf{S}[m] = (\mathbf{G}_m[p])\mathbf{S}[m]\mathbf{G}[p]$	$\mathbf{G}[p_1]\mathbf{G}[p_2] = \mathbf{G}[p_2 + p_1]$	$\mathbf{G}[p]\mathbf{S}[m] = (\mathbf{G}_m[p])\mathbf{S}[m]\mathbf{G}[p]$
S	$\mathbf{S}[m]\mathbf{V}[a] = \mathbf{V}[a]\mathbf{S}[am]$	$\mathbf{S}[m]\mathbf{F} = \mathbf{F}\mathbf{S}[m]$	$\mathbf{S}[m]\mathbf{Q}[c] = (\mathbf{Q}_m[c])\mathbf{Q}[c]\mathbf{G}[m]$	$\mathbf{S}[m]\mathbf{R}[d] = \mathbf{R}[d]\mathbf{S}[m]$	$\mathbf{S}[m]\mathbf{G}[p] = (\mathbf{G}_m[-p])$ $\times \mathbf{G}[p]\mathbf{S}[m]$	$\mathbf{S}[m_1]\mathbf{S}[m_2] = \mathbf{S}[m_2 + m_1]$

TABLE 5.1. Operator Algebra

5.10 REFERENCES

[1] M. Nazarathy and J.W. Goodman, Diffraction transforms in homogeneous birefringent media, *J. Opt. Soc. Am.* **A 3**, 523–531 (1986).

[2] M. Nazarathy and J. Shamir, First order optics —a canonical operator representation: lossless systems, *J. Opt. Soc. Am.* **72**, 356–364 (1982).

[3] M. Nazarathy and J. Shamir, First-order optics: operator representation for systems with loss or gain, *J. Opt. Soc. Am.* **72**, 1398–1408 (1982).

[4] R.L. Fork, O.E. Martínez, and J.P. Gordon, Negative dispersion using pairs of prisms, *Opt. Lett.* **9**, 150–152 (1984).

[5] J.P. Gordon and R.L. Fork, Optical resonator with negative dispersion, *Opt. Lett.* **9**, 153–155 (1984).

[6] L.D. Landau and E.M. Lifshitz, *Electrodynamics of Continuous Media* (Pergamon Press, Oxford, 1960).

[7] J.W. Goodman, *Introduction to Fourier Optics* (McGraw-Hill, New York, 1968).

[8] R.A. Athale, J.N. Lee, E.L. Robinson, and H.H. Szu, Acousto-optic processors for real-time generation of time-frequency representations, *Opt. Lett.* **8**, 166–168 (1983).

[9] M. Nazarathy and J.W. Goodman, Ambiguity function processors using linear birefringent dispersive materials, *Proceedings of SPIE, Spatial Light Modulators and Applications* **467** (1984).

6

Elements of Euclidean optics

Kurt Bernardo Wolf

ABSTRACT Euclidean optics are models of the manifold of rays and wavefronts in terms of coset spaces of the Euclidean group. One realization of this construction is the geometric model of Hamilton's optical phase space. Helmholtz optics is a second Euclidean model examined here. A wavization procedure is given to map the former one on the latter. Non-euclidean transformations of the manifold of rays are provided by Lorentz boosts that produce a global "4π" comatic aberration.

6.1 Introduction

> ***A plane is said to be similarly inclined to a plane as another is to another when the said angles of the inclinations are equal to one another... Equal and similar solid figures are those contained by similar planes equal in multitude and in magnitude.***
>
> Euclid, *Elements*
> Book XI, *Definitions 7 and 9*

This monograph structures the results of several previous articles, some written more than a decade ago, in the light of discussions held during and after the second Lie Optics workshop. The considerable advances in the use of Lie algebraic methods for magnetic and light optical design suggest their application to other closely related areas, such as polarization and wave optics, and invites incorporating more distant fields such as signal analysis and tomography, in the directions of polychromatic and far-metaxial optics.

The motivation of the present work may be focused in two questions that can be posed, in the context of their current solution, in the following terms:

1. *How far off the optical axis can we go?* Paraxial optics with its linear transformations of phase space is, of course, the starting point. The art of aberration expansions into the metaxial regime has been refined by their classification, computation, and understanding the way they compound in propagating through a system. The validity of such expansions must stop somewhere, however; probably much before 90°, and certainly before 180°.

2. *How do we wavize geometric optics?* Again, the paraxial regime may be said to be domestic territory: the symplectic group of canonical integral transforms provides a well-established bridge to transit from geometric to wave optics, as it does between classical and quantum mechanics in systems with quadratic potentials. On the wave side, it couples naturally with Fourier and coherent-state optics. Hilbert spaces, Wigner distribution functions, and measurement theories, often translated from quantum mechanics, are available. The metaxial regime does not seem to have such a reliable wavization procedure. In particular, we would like to be able to design an optical system with the tools of Lie geometric optics, and thereby know its behavior as we turn on the wave nature of light. Lastly, there is a gulf in the *global* regime, *i.e.*, "4π"optics that extends over the *full* sphere of ray directions.

From the experience of quantum mechanics, it is evident that both questions are related. There, *global* properties of the potential (over the full real line or 3-space) are most important, and the 'far-away' regions can be seldom ignored.[1] Yet, the most striking *difference* between (nonrelativistic point-particle) mechanics and (geometric) optics is in their phase spaces: the former is flat and unbounded both in position and in momentum, while in the latter momentum ranges over a sphere projected flat on two disks in its equatorial plane. The Heisenberg-Weyl group of phase space motions underlies the symplectic geometry of the former, but not of the latter, unless we replace optics by its paraxial regime. It is our contention that the basic group of *global* optics is the three-dimensional *Euclidean* group $\mathcal{E}_3 = \mathrm{ISO}(3)$ of rigid motions of three-space.

In Section 2 we examine the structure of the manifold of rays of geometric optics, a vector bundle $\wp$, introducing the local and standard screen coordinates, and the *Descartes* sphere of ray directions. Section 3 reviews the composition rule for the Euclidean group of Lie operators that take a standard frame to any position and orientation. Thus we have a Euclidean theory of *frames*. The infinitesimal *generators* yield the Euclidean algebra on that group manifold given in Section 4.

When the objects that we regard as elementary have a *symmetry* group $\mathcal{H} \subset \mathcal{E}_3$, a corresponding model of optics follows. For example, in geometric optics rays are pictured as straight lines filling space, with a $\mathcal{T}_1$ symmetry under translations along the line times $\mathcal{R}_2$ under rotations around it. Sections 5, 6, and 7 show that the manifold of rays $\wp$ is the space of *cosets* $\mathcal{H}\backslash\mathcal{E}_3$. The Euclidean group and algebra are realized on that space, and shown to be canonical in the usual symplectic sense of Hamilton's theory, here derived from the conservation of the Haar measure. Indeed, also po-

[1]On the other hand, in quantum mechanics, kinetic energy is mostly of the fixed standard form $p^2/2m$.

larization and signal optics may be identified with coset spaces, $\mathcal{R}_2\backslash\mathcal{E}_3$ and $\mathcal{T}_1\backslash\mathcal{E}_3$, respectively.

Our second main model of interest is *wave* optics. When the elementary object is a plane, it defines *wavefront* optics.[2] This we introduce in Section 8 as the manifold $\mathcal{E}_2\backslash\mathcal{E}_3$, and note that it can carry *signals* in a train amenable to Fourier analysis. In Section 9 we show that the Fourier decomposition is irreducible under Euclidean transformations, each wavenumber component satisfying a corresponding *Helmholtz* equation that is the Casimir invariant of the algebra. Following the *modus operandi* of quantum mechanics, Section 10 builds the Hilbert space of oscillatory solutions over the standard screen whose inner product, uniquely invariant under Euclidean motion and endowed with a non-local measure, was previously found with Stanly Steinberg (Albuquerque). This, rather than $\mathcal{L}_2(\Re^2)$, seems to be the Hilbert space appropriate for wavized optics because, as shown in Section 11, it involves by necessity and on the same footing, both the wave function and its *normal derivative* at the screen.[3] With these elements we propose a definite "4π wavization" process that leads from geometric to wave optics on the level of the Euclidean group.

Up to here, we deal only with a group theory of rigid motions. Section 12 introduces the *Lorentz* transformation responsible for stellar aberration. Although this phenomenon has been known for centuries, its implications for Hamiltonian optics in image formation seem to have been overlooked. Sections 13 and 14 contain the results of two recent papers with Natig Atakishiyev (Baku) and Wolfgang Lassner (Leipzig) that predict a global *comatic aberration* of geometric and wave images on boosted screens, stemming from the nonlinear action of the Lorentz group on the corresponding coset space models. The proposed wavization process is applied to these transformations, and seen to hold.

Field theories on groups in empty space have been abundant —and some have been very important; yet optics visibly needs the dynamics of inhomogeneous media. We have felt obliged thus to add some preliminary reflections on refraction among the concluding remarks in Section 15. This process appears as a *coupling* between representations of the Euclidean group through the conservation law due to Willebrord Snell (experimentally) and René Descartes (theoretically). We are inspired by the Cartesian *Mèthode* in regarding Optics as Nature observing Symmetry, because in that way it pleases the mind.

---

[2]When only $\mathcal{T}_2$ symmetry is present, it describes *polarized* polichromatic wave optics.

[3]Quantum mechanics uses only the first one because the Schrödinger equation has a first degree derivative only in the evolution variable.

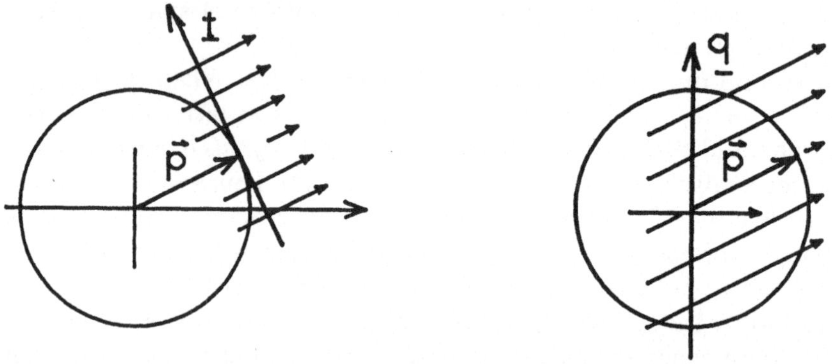

FIGURE 1. (*a*) Local screen parametrization of a bundle of parallel rays. (*b*) Parametrization by a fixed reference screen.

6.2 The bundle of rays in geometric optics

> *We ought to give the whole of our attention to the most insignificant and most easily mastered facts, and remain a long time in contemplation of them until we are accustomed to behold the truth very clearly and distinctly.*
>
> René Descartes, *Rules for the Direction of the Mind*
> Rule IX

In geometric optics light rays are modeled as lines in 3-space oriented in all directions. In a homogeneous medium the lines are straight, and usual Euclidean geometry applies. We examine here this manifold $\wp$ of oriented lines to show its structure, and find good sets of coordinates.

With the tools of thought we can sort out all those lines that are oriented in a chosen direction and identify them by a point $\vec{n}$ on a sphere $\mathcal{S}_2$. This is a projection $\pi : \wp \mapsto \mathcal{S}_2$. The inverse image $\pi^{-1}(\vec{n})$ of such a point is a set of paralell rays that can be brought one onto the other by translations within their perpendicular plane. The set of translations constitute a group and a vector space $\mathcal{T}_2$. These properties of structure are those of a *vector bundle* [1]. The manifold of light rays in geometric optics is thus a (non-trivial) vector bundle $\wp$, with an $\mathcal{S}_2$ *base space* of ray directions, and a $\mathcal{T}_2$ *local screen*, the typical *fiber* of the bundle. The screens are local because each is associated to its set of parallel rays and perpendicular to them. Two rays that are not paralell will have their screens oriented differently. Let us now introduce coordinates.

The set of directions of the rays —the $\mathcal{S}_2$ base space of the bundle— is the Descartes sphere. This can be parametrized by a pair of usual Euler angles $\{\theta, \phi\}$, with the well-known but treatable difficulties of coordinates on spheres. Alternatively, we may choose a Cartesian three-vector $\vec{p} = (p_x, p_y, p_z)$ on a sphere of fixed radius $n = |\vec{n}| = \sqrt{p_x^2 + p_y^2 + p_z^2}$. (Of course,

n will be the refractive index of the medium!) Two of the three coordinates, say $\mathbf{p} = (p_x, p_y)$, supplemented by the sign σ of the otherwise redundant third component $p_z = \sigma\sqrt{n^2 - p_x^2 - p_y^2}$, will also serve to indicate a point of the Descartes sphere uniquely.[4] Depending on the need we shall use one or the other parametrization for $\mathcal{S}_2$.

The local screen of each paralell pencil of rays is an $\Re^2$ Cartesian manifold of translation two-vectors $\{t_{x'}, t_{y'}\}$ with an origin or *optical center* and an orientation of its local x' and y' axes. We may choose these in the direction of the p_x and p_y axes of the Descartes sphere, but we will have orientation trouble when effecting the paralell transport of the screens around the sphere.[5] Such features are normal in bundles that are not direct products, but we may always work on local charts in a sizable neighborhood of some standard ray. See Figure 1(a).

A local parametrization that is preferred in Hamiltonian optics is that of a *standard screen*. This screen is referred to an optical axis $\vec{p}_0 = (0, n)$, placed at $z = 0$. Rays within paralell pencils are parametrized by their observable of *position*, *i.e.*, their intersection $\mathbf{q} = (q_x, q_y)$ with that standard screen. See Figure 1(b). This works well except for rays that are paralell to the screen, all of which will map on a point at infinity. Again, the coordinates parametrize well sizable neighborhoods of rays, but cannot be global. Since it *does* happen that $\mathbf{q}$ and $\mathbf{p}$ are *canonically conjugate* in every neighborhood, they do deserve particular attention.

We may approach similarly other models, such as the manifold of frames (oriented point-objects), ribbons or of screws for polarization optics, and planes (for wavefront optics). We see naturally such manifolds as coset spaces of the group of Euclidean motions, modulo the symmetry subgroup of the object.

6.3 Lie operators on the Euclidean group

The action of the elements g of a Lie group $\mathcal{G}$ on the space of functions f of its own manifold variables γ [2], is carried by Lie transformations $\mathcal{L}_g$ [3] that may be realized in at least two ways: through right or left action, *viz.*,

$$\mathcal{L}_g^{\mathrm{R}} f(\gamma) = f(\mathcal{L}_g^{\mathrm{R}}\gamma) = f(\gamma g), \tag{3.1a}$$

$$\mathcal{L}_g^{\mathrm{L}} f(\gamma) = f(\mathcal{L}_g^{\mathrm{L}}\gamma) = f(g^{-1}\gamma). \tag{3.1b}$$

---

[4]When $p_z = 0$ we may take $\sigma = 0$.

[5]The rays of geometric optics may be easily parametrized in Cartesian solid geometry by the line $\vec{r}(s) = s\vec{p} + \vec{t}$, where $s \in \Re$ measures length along the ray in units of $1/n$. The vector $\vec{t}$ may be chosen orthogonal to $\vec{p}$, namely $\vec{t} \cdot \vec{p} = 0$, so it contains only two independent local screen parameters.

Still another one is the conjugation $\mathcal{L}_g^{\mathrm{C}} f(\gamma) = f(g^{-1}\gamma g)$, and in fact any subgroup $\mathcal{G}$ of $\mathcal{G} \times \mathcal{G}$ may be used with bilateral action [4].

Lie transformations may be generally represented as exponentials of differential operators of first degree in the manifold coordinates of γ. They have the above property of "jumping into" the function's arguments. Let us verify step by step that the right action (3.1*a*) is consistent with the requirement that the group composition property be preserved, *i.e.*, that $\mathcal{L}_{g_1}^{\mathrm{R}} \mathcal{L}_{g_2}^{\mathrm{R}} = \mathcal{L}_{g_1 g_2}^{\mathrm{R}}$ holds when acting on any function f. We use (3.1*a*) for the rightmost factor to write

$$\mathcal{L}_{g_1}^{\mathrm{R}} \mathcal{L}_{g_2}^{\mathrm{R}} f(\gamma) = \mathcal{L}_{g_1}^{\mathrm{R}} f(\mathcal{L}_{g_2}^{\mathrm{R}} \gamma) = \mathcal{L}_{g_1}^{\mathrm{R}} f(\gamma g_2), \tag{3.2a}$$

and call

$$f_2(\kappa) = f(\mathcal{L}_{g_2}^{\mathrm{R}} \kappa) = f(\kappa g_2), \tag{3.2b}$$

for any κ such as γ. Thus we continue (3.2*a*) writing

$$\mathcal{L}_{g_1}^{\mathrm{R}} f_2(\gamma) = f_2(\mathcal{L}_{g_1}^{\mathrm{R}} \gamma) = f_2(\gamma g_1). \tag{3.2c}$$

Now we can use (3.2*b*) again, with $\kappa = \gamma g_1$, and finish (3.2*a*) with

$$\mathcal{L}_{g_1}^{\mathrm{R}} \mathcal{L}_{g_2}^{\mathrm{R}} f(\gamma) = f(\gamma g_1 g_2) = \mathcal{L}_{g_1 g_2}^{\mathrm{R}} f(\gamma). \tag{3.2d}$$

Observe carefully that $\mathcal{L}_{g_1}^{\mathrm{R}}$ acts *first* and $\mathcal{L}_{g_2}^{\mathrm{R}}$ *second* on γ to yield $\gamma g_1 g_2$, as read in the usual direction.[6]

The Euclidean group $\mathcal{E}_3 = \mathrm{ISO}(3)$ contains the group of rotations $\mathcal{R}_3 = \mathrm{SO}(3)$ and of translations $\mathcal{T}_3$. It has the following well-known structure that we display in coordinates. Let $\mathcal{R}_3 \ni \mathbf{R}$ be an orthogonal 3×3 matrix, of unit determinant, and $\mathcal{T}_3 \ni \vec{v} = (v_x, v_y, v_z)$ a Cartesian three-dimensional row-vector.[7] We may denote the elements of the Euclidean group $\mathcal{E}_3$ as

$$\mathsf{E}(\mathbf{R}, \vec{v}) = \mathcal{L}_{\mathbf{R}}^{\mathrm{R}} \mathcal{L}_{\vec{v}}^{\mathrm{R}} = \mathsf{E}(\mathbf{R}, \vec{0})\mathsf{E}(\mathbf{1}, \vec{v}). \tag{3.3a}$$

The two subgroup products are denoted by matrix multiplication and (commutative) sum as

$$\mathsf{E}(\mathbf{R}_1, \vec{0})\mathsf{E}(\mathbf{R}_2, \vec{0}) = \mathsf{E}(\mathbf{R}_1\mathbf{R}_2, \vec{0}), \tag{3.3b}$$

$$\mathsf{E}(\mathbf{1}, \vec{v}_1)\mathsf{E}(\mathbf{1}, \vec{v}_2) = \mathsf{E}(\mathbf{1}, \vec{v}_1 + \vec{v}_2). \tag{3.3c}$$

The yuxtaposition of the two subgroups is that of *semidirect product*, $\mathcal{E}_3 = \mathcal{R}_3 \triangleright \mathcal{T}_3$, specified by the action of $\mathcal{R}_3$ on $\mathcal{T}_3$, in row-vector and matrix notation

$$\mathsf{E}(\mathbf{R}, \vec{0})\mathsf{E}(\mathbf{1}, \vec{v})\mathsf{E}(\mathbf{R}^{-1}, \vec{0}) = \mathsf{E}(\mathbf{1}, \vec{v}\mathbf{R}^{-1}). \tag{3.3d}$$

---

[6] A similar argument verifies that (3.1*b*), *i.e.*, action from the left by $g^{-1}\gamma$ is also consistent with the group property. *Not* consistent with it would be $g\gamma$, γg^{-1}, $g\gamma g$, etc.

[7] We use the top arrow $\vec{\ }$ to denote three-dimensional row-vectors.

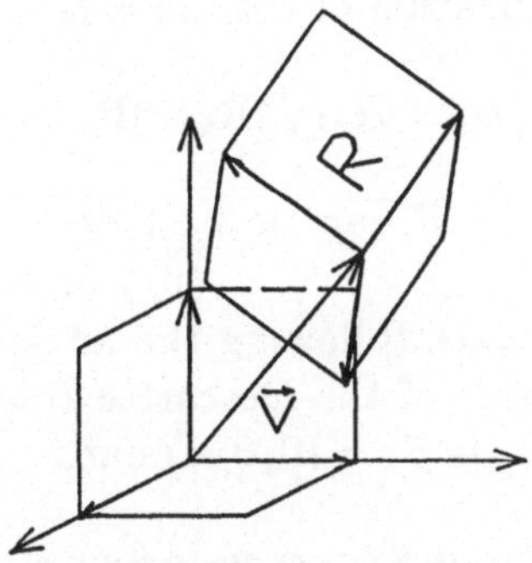

FIGURE 2. The standard and a Euclidean-displaced frame.

The translation subgroup is thus the invariant or *normal* subgroup of the Euclidean group. From here, the group multiplication law is

$$\mathsf{E}(\mathbf{R}_1, \vec{v}_1)\mathsf{E}(\mathbf{R}_2, \vec{v}_2) = \mathsf{E}(\mathbf{R}_1\mathbf{R}_2, \vec{v}_1\mathbf{R}_2 + \vec{v}_2). \tag{3.3e}$$

The group unit is $\mathsf{E}(\mathbf{1}, \vec{0}\,)$, and the inverse is $\mathsf{E}(\mathbf{R}, \vec{v}\,)^{-1} = \mathsf{E}(\mathbf{R}^{-1}, -\vec{v}\,\mathbf{R}^{-1})$.

Functions on the six-dimensional $\mathcal{E}_3$ manifold will be denoted by $f(\mathbf{P}, \vec{r}\,)$, and we shall refer generically to $\mathbf{P}$ as *direction* and to $\vec{r}$ as *position*, because of ulterior motives. Direction is subject to rotation, and position to both rotation and translation. We act on these functions first with the right Lie transformation of the rotation $\mathcal{L}^{\mathrm{R}}_{\mathbf{R}}$ and second with that of the translation $\mathcal{L}^{\mathrm{R}}_{\vec{v}}$, to get

$$\mathsf{E}(\mathbf{R}, \vec{v}\,)f(\mathbf{P}, \vec{r}\,) = \mathcal{L}^{\mathrm{R}}_{\mathbf{R}}\mathcal{L}^{\mathrm{R}}_{\vec{v}}f(\mathbf{P}, \vec{r}\,) = \mathcal{L}^{\mathrm{R}}_{\mathbf{R}}f(\mathbf{P}, \vec{r} + \vec{v}\,) = f(\mathbf{P}\mathbf{R}, \vec{r}\,\mathbf{R} + \vec{v}\,), \tag{3.4}$$

by (3.1*a*) and (3.3*e*). Upon this, the group unit $\mathsf{E}(\mathbf{1}, \vec{0}\,)$ is rotated by $\mathbf{R}$ and translated by $\vec{v}$ to $\mathsf{E}(\mathbf{R}, \vec{v}\,)$. The group unit may be seen as a *standard frame* of three orthogonal axes, defining the origin both of direction and position. See Figure 2. Through rotations and translations, we may bring this frame to any other one, mounted on the generic $\vec{r}$ and with the orientation obtained by the action of the rotation matrix $\mathbf{P}$ on the standard frame. At this stage we may say that we are defining the model of frames as the set of $\mathcal{R}_3$–orientable objects in $\mathcal{T}_3$–space. Note that the translation by $\vec{v}$ of the frames' origin is performed with reference to the *standard* frame, *i.e.*, $\vec{r}\,\mathbf{R} + \vec{v}$, and not some $(\vec{r} + \vec{v}\,)\mathbf{R}$ appears in (3.4). This corresponds to the preferred parametrization of rays in optics by a *fixed* standard screen[8]. Lie operators act moving the underlying space referred to a fixed frame.

[8]On the other hand, the natural bundle parametrization by *local* screens discussed in the previous Section proceeds in accordance with the *left* action of Lie operators in (3.1*b*). For these, the parametrization of $\mathcal{E}_3$ by (3.3*a*) may be

The Euler angle parametrization of rotations is

$$\mathbf{R}(\psi,\theta,\phi) = \mathbf{R}_z(\psi)\mathbf{R}_y(\theta)\mathbf{R}_z(\phi). \tag{3.5a}$$

The effect of a rotation $\mathbf{R}$ on $\mathbf{P}$ can be best des by letting $\mathbf{R}$ act on at least two distinct row-vectors $\vec{p} = (p_x, p_y, p_z)$ of length n, and referring to the z–axis as the *optical axis*, following the ancient convention used by opticists. It is the *forward pole* of the Descartes sphere. Thus $\mathbf{R}$ will map the direction of the optical axis $\vec{p}_o = (0,0,n)$ onto

$$\vec{p} = \vec{p}_o\mathbf{R}(\psi,\theta,\phi) = (n\sin\theta\cos\phi, n\sin\theta\sin\phi, n\cos\theta). \tag{3.5b}$$

We note that the angle ψ is absent from the right-hand side of the last equation. Similar formulae can be produced for the other directions, say $(n,0,0)$ (that *will* contain ψ). The action of $\mathbf{R}$ on the whole $\vec{p}$–sphere is *transitive* (a right frame may be mapped to any other right frame) and *effective* (no frame is left invariant except by the group identity).

6.4 Generators of the Euclidean group

As a standard assumption in Lie theory, our group $\mathcal{G}$ has its manifold parametrized by a set of coordinates, $g(x)$, $x = \{x_i\}_{i=1}^N$, $g(0) =$ identity, and for vanishing ϵ we can write $\mathcal{L}_{g(\epsilon)}f(\gamma) = f(\gamma) + \epsilon_i\hat{\Gamma}_i f(\gamma) + \cdots$. The $\hat{\Gamma}_i$ are the *generators* of the group and are *realized* as first-order differential operators in the coordinates of γ [2]. The set of these generators $\{\hat{\Gamma}_i\}$ closes under commutation into a Lie *algebra* $[\hat{\Gamma}_i, \hat{\Gamma}_j] = c_{ijk}\hat{\Gamma}_k$, and the Lie transformations may be written as $\mathcal{L}_{g(x)} = \exp x_i\hat{\Gamma}_i$.

We may find the generators of the Euclidean group on functions of its own six-dimensional manifold $f(\mathsf{E}(\mathbf{P},\vec{r})) = f(\mathbf{P},\vec{r})$ from (3.4); these will constitute the *Euclidean* Lie algebra in the *frame* realization. The right

---

conveniently replaced by

$$\mathsf{E}^l(\vec{v},\mathbf{R}) = \mathcal{L}^{\mathrm{L}}_{\vec{v}}\mathcal{L}^{\mathrm{L}}_{\mathbf{R}} = \mathsf{E}(\mathbf{R},\vec{v}\mathbf{R}).$$

Accordingly, the action on the space of frames now parametrized accordingly is

$$\mathsf{E}^l(\vec{v},\mathbf{R})f(\vec{t},\mathbf{P}) = f([\vec{t}-\vec{v}]\mathbf{R}^{-1}, \mathbf{R}^{-1}\mathbf{P}).$$

Here, the translation $\vec{v}$ is in the frame of $\vec{t}$ as it is rotated by $\mathbf{R}^{-1}$. Parallel developments can be made for left group action $g^{-1}\gamma$ and their Lie operators. The third possibility, $g^{-1}\gamma g$, favors parametrization by *conjugation classes*. For the $\mathcal{R}_3$ subgroup, this is through specifying the rotation axis $\hat{n}(\theta,\phi)$ and the rotation angle χ around that axis. The latter labels the classes. Under the adjoint action of $\mathcal{R}_3$, $\hat{n}$ transforms as a vector and χ is invariant.

translations $\mathcal{E}_{\vec{v}}^{\mathrm{R}} = \exp \vec{v} \cdot \hat{\vec{T}}$ act as

$$\exp \vec{v} \cdot \hat{\vec{T}} f(\mathbf{P}, \vec{r}) = f(\mathbf{P}, \vec{r} + \vec{v}). \tag{4.1a}$$

We thus find

$$\hat{T}_x = \frac{\partial}{\partial r_x}, \qquad \hat{T}_y = \frac{\partial}{\partial r_y}, \qquad \hat{T}_z = \frac{\partial}{\partial r_z}. \tag{4.1b}$$

Translations do not affect ray direction, so no p_i–derivatives appear.

For rotations, we may use the explicit expressions of the 3×3 matrices that act in (3.4) both the $\vec{p} = (p_x, p_y, p_z)$ and $\vec{r} = (r_x, r_y, r_z)$ row-vectors, that transform in unison. We call $\hat{R}_i$ the generators of the finite rotation matrices $\mathbf{R}_i$, in the following way:

$$\exp \alpha_x \hat{R}_x \quad \mapsto \quad \mathbf{R}_x(\alpha_x) = \begin{pmatrix} 1 & 0 & 0 \\ 0 & \cos\alpha_x & \sin\alpha_x \\ 0 & -\sin\alpha_x & \cos\alpha_x \end{pmatrix}, \tag{4.2a}$$

$$\exp \alpha_y \hat{R}_y \quad \mapsto \quad \mathbf{R}_y(\alpha_y) = \begin{pmatrix} \cos\alpha_y & 0 & -\sin\alpha_y \\ 0 & 1 & 0 \\ \sin\alpha_y & 0 & \cos\alpha_y \end{pmatrix}, \tag{4.2b}$$

$$\exp \alpha_z \hat{R}_z \quad \mapsto \quad \mathbf{R}_z(\alpha_z) = \begin{pmatrix} \cos\alpha_z & \sin\alpha_z & 0 \\ -\sin\alpha_z & \cos\alpha_z & 0 \\ 0 & 0 & 1 \end{pmatrix}. \tag{4.2c}$$

The rotation generators take their very familiar form

$$\hat{R}_x = p_y \frac{\partial}{\partial p_z} - p_z \frac{\partial}{\partial p_y} + r_y \frac{\partial}{\partial r_z} - r_z \frac{\partial}{\partial r_y}, \tag{4.3a}$$

$$\hat{R}_y = p_z \frac{\partial}{\partial p_x} - p_x \frac{\partial}{\partial p_z} + r_z \frac{\partial}{\partial r_x} - r_x \frac{\partial}{\partial r_z}, \tag{4.3b}$$

$$\hat{R}_z = p_x \frac{\partial}{\partial p_y} - p_y \frac{\partial}{\partial p_x} + r_x \frac{\partial}{\partial r_y} - r_y \frac{\partial}{\partial r_x}. \tag{4.3c}$$

The length n of the vector $\vec{p}$, the radius of the Descartes sphere, is naturally a Euclidean invariant in homogeneous media —only.

The hatted operators are a vector basis for the Lie algebra. As we can check, their *commutators* $[\hat{X}, \hat{Y}] = \hat{X}\hat{Y} - \hat{Y}\hat{X}$ close:

$$[\hat{T}_i, \hat{T}_j] = 0, \qquad i, j = x, y, z \tag{4.4a}$$

$$[\hat{R}_i, \hat{T}_j] = -\varepsilon_{ijk} \hat{T}_k, \qquad \varepsilon_{xyz} = 1, \tag{4.4b}$$

$$[\hat{R}_i, \hat{R}_j] = -\varepsilon_{ijk} \hat{R}_k, \tag{4.4c}$$

and all other independent commutators are zero. This clearly displays the Euclidean algebra as the *semi-direct sum* of the translation and rotation subalgebras [2].[9]

---

[9] At this point we could introduce rotations that act exclusively on the frame

6.5 Coset spaces and rays

The realization of frames seen in the last Section will serve now to define other objects subject to Euclidean group action. In scalar geometric optics, light rays are modeled by straight lines, with no particular origin nor polarization plane. If such an object is rotated around its axis or translated in its direction, it is still the same elementary object. These are *symmetry transformations* of the object, and they always form a *group* [2]. In the geometric optics case, the group is $\mathcal{H}^{\rm geom} = \mathcal{R}_2 \times \mathcal{T}_1$, the direct product of the rotation group in two dimensions with the translation group in one dimension. The *rays* of a model will be identified thus as the *equivalence classes* of the Euclidean group modulo the symmetry subgroup of the object. We will now formalize this construction presenting some standard material on the equivalence classes in a group called *cosets*, before we apply it to the Euclidean group.

If $\mathcal{G}$ is a group and $\mathcal{H}$ a subgroup, we may divide the manifold of $\mathcal{G}$ into disjoint subsets by $\mathcal{H}$ in the following way: let $g \in \mathcal{G}$ and consider the set $\{hg\}_{h\in\mathcal{H}}$, called the (left) *coset* of g by $\mathcal{H}$. We thereby introduce the relation $g_1 \equiv g_2$ between two elements of $\mathcal{G}$ when $\mathcal{H}g_1 = \mathcal{H}g_2$. The coset of the identity $e \in \mathcal{G}$ is $\mathcal{H}e = \mathcal{H}$. Clearly, the cosets of g and of hg are equal, and from here it is easily shown that $\equiv$ is an *equivalence* relation, *i.e.*, it divides the manifold of $\mathcal{G}$ into disjoint subsets; the *set* of left cosets is denoted by $\mathcal{H}\backslash\mathcal{G}$. Within every coset we may choose a *representative* element $\gamma \in \mathcal{G}$. We can thus display the structure of $\mathcal{G}$ to be that of a *fiber* bundle whose base space is the set of these representatives γ, namely $\mathcal{H}\backslash\mathcal{G}$, and whose typical fiber is $\mathcal{H}$. The projection operator $\pi{:}\mathcal{G} \mapsto \mathcal{H}\backslash\mathcal{G}$ may be used to introduce subgroup-adapted coordinates on the manifold of $\mathcal{G}$ by writing g as $g(\rho, v) = h(\rho)\gamma(v)$, with h parametrized by coordinates ρ for $\mathcal{H}$, and coordinates v for $\gamma \in \mathcal{H}\backslash\mathcal{G}$ as representative. It may be that in a badly chosen set of representatives, some elements of will not admit such a decomposition; in approaching these elements, some of its coordinates in v will escape to infinity.

Left cosets transform under *right* group action. If under $k \in \mathcal{G}$ the group manifold transforms as $g \mapsto g' = \mathcal{L}_k^{\rm R}{:}g = gk$, and the subgroup–coset coordinates read $g(\rho, v) = h(\rho)\gamma(v)$ and $g'(\rho, v) = g(\rho', v') = h'(\rho)\gamma'(v) = h(\rho')\gamma(v')$, then we may subduce the mapping $f(v) \mapsto f'(v) = f(v') = \mathcal{L}_k^{{\rm R},\mathcal{H}}{:}f(v)$, where $\mathcal{L}_k^{{\rm R},\mathcal{H}}$ is now a Lie transformation acting on the functions f of the coset space coordinates v.

For the Euclidean group $\mathcal{E}_3 = \mathcal{R}_3 \triangleright \mathcal{T}_3$, we have that $\mathcal{R}_3\backslash\mathcal{E}_3 = \mathcal{T}_3$ and

---

orientation $\vec{p}$-space, as $\hat{S}_x = p_y\partial_{p_z} - p_z\partial_{p_y}$, etc., with commutators $[\hat{S}_i, \hat{T}_j] = 0$, $[\hat{S}_i, \hat{R}_j] = -\hat{S}_k$, $[\hat{S}_i, \hat{S}_j] = -\hat{S}_k$. The structure of this group would be $\mathcal{R}_3^{\rm S} \times (\mathcal{R}_3^{\rm R-S} \triangleright \mathcal{T}_3)$.

$\mathcal{E}_3/\mathcal{T}_3 = \mathcal{R}_3$ are groups themselves and have been used for the global parametrization of $\mathsf{E}(\mathbf{R}, \vec{v})$.[10] We will now show that geometric optics is a model based on the $\wp = (\mathcal{R}_2 \times \mathcal{T}_1)\backslash\mathcal{E}_3$ manifold. Let us now see the way the coordinates of $\mathcal{R}_2\backslash\mathcal{E}_3$ and $\mathcal{T}_1\backslash\mathcal{E}_3$ transform under Euclidean action.

For the rotation group $\mathcal{R}_3 \subset \mathcal{E}_3$ the Euler angle parametrization (3.5) is the appropriate one when the symmetry group $\mathcal{R}_2$ is $\mathbf{R}_z(\psi)$ with $\psi \in \mathcal{S}_1$. Each coset is thus a one-dimensional sphere, *i.e.*, a circle. The cosets $\mathcal{R}_2\mathbf{R}(\psi,\theta,\phi) = \mathcal{R}_2\mathbf{R}(0,\theta,\phi)$ are the points of the space $\mathcal{R}_2\backslash\mathcal{R}_3$, and the appropriate coset *representatives* are plainly $\gamma(\theta,\phi) = \mathbf{R}(0,\theta,\phi)$. Their manifold is the two-sphere: $\mathcal{R}_2\backslash\mathcal{R}_3 = \mathcal{S}_2$. Transformations of the points of this sphere by $\mathcal{L}^{\mathrm{R}}_{\mathbf{S}}$ under a rotation $\mathbf{S}(\alpha,\beta,\gamma)$ are found, as usual, by simply applying the matrix $\mathbf{S}$ to the row vector in (3.5*b*). The result will yield $f'(\theta,\phi) = f(\theta',\phi')$, $\theta'(\theta,\phi;\alpha,\beta,\gamma)$ and $\phi'(\theta,\phi;\alpha,\beta,\gamma)$. The manifold of cosets $\mathcal{R}_2\backslash\mathcal{E}_3$ is thus parametrized by $\{\gamma(\theta,\phi),(r_x,r_y,r_z)\} = \{\vec{p},\vec{r}\}$. The model $\mathcal{R}_2\backslash\mathcal{E}_3$ describes objects that are points in space $\vec{r}$ with a direction vector $\vec{p}$ of fixed length on each point, *i.e.*, a special kind of *vector field*. It has one dimension less than the original six-dimensional full group manifold.

We examine now the cosets by the *translation* symmetry subgroup $\mathcal{T}_1$ along the z-axis. These are built in an analogous way, but with the difference that while $\mathcal{T}_3$ and its subgroups splits off easily to the *right* of $\mathcal{E}_3 = \mathcal{R}_3 \triangleright \mathcal{T}_3$, we need it to the *left*. The standard frame $e = \mathsf{E}(\mathbf{1},(0,0,0))$ and the line of z-translated frames $\mathsf{E}(\mathbf{1},(0,0,s))$, $s \in \Re$, constituting the $\mathcal{T}_1$ coset of the identity, parametrize the same standard ray in the geometric optics model. Similarly, an arbitrary frame $\mathsf{E}(\mathbf{P},\vec{r})$ and the coset of frames $\mathsf{E}(\mathbf{1},(0,0,s))\mathsf{E}(\mathbf{P},\vec{r}) = \mathsf{E}(\mathbf{P},\vec{r} + (0,0,s)\mathbf{P})$, $s \in \Re$ describe the same ray, whose position vector is, in coordinates

$$(r_x,r_y,r_z) + (0,0,s)\mathbf{P} = (r_x + s\sin\theta\cos\phi, r_y + s\sin\theta\sin\phi, r_z + s\cos\theta), \tag{5.1}$$

and carries the orientation $\mathbf{P}(\psi,\theta,\phi)$. Further cosetting by $\mathcal{R}_2$ will elliminate the plane polarization angle ψ.

The space of cosets $\mathcal{H}^{\mathrm{geom}}\backslash\mathcal{E}_3$, $\mathcal{H}^{\mathrm{geom}} = \mathcal{R}_2 \times \mathcal{T}_1$ we shall show now, is the manifold of geometric optics rays $\wp$ described in Section 2. The elements of $\mathcal{H}^{\mathrm{geom}}$ are $\mathsf{E}(\mathbf{R}_z(\psi),(0,0,s))$, $\psi \in \mathcal{S}_1$, $s \in \Re$, a cylindrical submanifold of $\mathcal{E}_3$. Every other coset is a right-translated version of this submanifold by $\mathsf{E}(\mathbf{P},\vec{r})$. The coset *representatives* may now be chosen in writing the decomposition

$$\mathsf{E}(\mathbf{P}(\psi,\theta,\phi),(r_x,r_y,r_z)) = \mathsf{E}(\mathbf{R}(\psi,0,0),(0,0,s))\ \mathsf{E}(\mathbf{P}(0,\theta,\phi),(q_x,q_y,0)), \tag{5.2}$$

where the vector $(q_x,q_y,0) = \vec{q}$ indicates a point *on the screen*, whose

[10]We may use $\mathcal{T}_3\backslash\mathcal{E}_3$ and $\mathcal{E}_3/\mathcal{R}_3$ for the *left* parametrization of $\mathsf{E}^l(\vec{t},\mathbf{R})$, corresponding to the coordinates of $\wp$ by local screens.

z–component will be henceforth assumed to be always *zero*. Thus we write

$$\vec{r} = (r_x, r_y, r_z) = (0,0,s)\mathbf{P}(0,\theta,\phi) + (q_x, q_y, 0) = s\vec{p}/n + \vec{q}. \qquad (5.3)$$

The coordinates of each $\mathcal{H}^{\text{geom}}$ coset are $\{\psi, s\}$ and those of the space of coset representatives are the four independent parameters in $\{\vec{p},\vec{q}\}$, where $\vec{p}$ lies on the Descartes sphere of radius n, and $\vec{q}$ indicates the ray intersection at the standard screen. Any nonzero value of ψ and r_z will fall into the $\mathcal{H}^{\text{geom}}$ factor to the left. Both are (by definition) unobservable in scalar geometric optics, but may be retained in the less stringent models of geometric optics with a polarization orientation or signals along the line.

We relate the group and coset parameters through

$$\begin{aligned} r_x &= q_x + s\sin\theta\cos\phi, & q_x &= r_x - r_z p_x/p_z, & (5.4a)\\ r_y &= q_y + s\sin\theta\sin\phi, \quad \textit{i.e.} & q_y &= r_y - r_z p_y/p_z, & (5.4b)\\ r_z &= 0 + sn\cos\theta, & s &= r_z\sec\theta = nr_z/p_z. & (5.4c)\end{aligned}$$

There is a submanifold where these coordinates fail, however: rays parallel to the standard screen —as may have been expected. For $\theta = \pi/2, p_z = 0$, and both s and $|\vec{q}|$ go to infinity because the decomposition (5.2) is impossible there. Figure 3 shows the geometrical situation in two dimensions. Comparison with Figure 1(*b*) justifies the identification $\wp = \mathcal{H}^{\text{geom}}\backslash\mathcal{E}_3$.

6.6 Euclidean group action on rays in geometric optics

The Euclidean transformations $\mathsf{E}(\mathbf{R}(\alpha,\beta,\gamma),\vec{v})$ of the rays in $\wp$ can now be found by acting from the right on $\mathsf{E}(\mathbf{P},\vec{r}) = h(\psi,s)\gamma(\vec{p}(\theta,\phi),(q_x,q_y))$[11] and then decomposing the product. See:

$$\mathsf{E}(\mathbf{P},\vec{r})\mathsf{E}(\mathbf{R},\vec{v}) = \mathsf{E}(\mathbf{PR},\vec{r}\mathbf{R}+\vec{v}) = h(\psi',s')\gamma(\vec{p}(\theta',\phi'),\vec{q}'). \qquad (6.1)$$

We use (5.2) for $\mathbf{P}' = \mathbf{PR}$, with $\mathbf{R}$–rotated angles $\{\psi',\theta',\phi'\}$ in place of $\{\psi,\theta,\phi\}$,

$$\vec{p}'(\theta,\phi) = \vec{p}(\theta',\phi') = \vec{p}(\theta,\phi)\mathbf{R}(\alpha,\beta,\gamma). \qquad (6.2)$$

This takes care of the direction part. Next, we look at the ray length parameter s' and the position coordinates $\vec{q}'$ on the screen in (5.4) with $\vec{r}' = \vec{r}\mathbf{R}(\alpha,\beta,\gamma)+\vec{v}$ in place of $\vec{r}$. This allows us to determine $s' = nr'_z/p'_z$ and

$$\vec{q}' = \vec{r}' - s'\vec{p}'/n = \vec{r}\mathbf{R} + \vec{v} - s'\vec{p}\mathbf{R}/n \qquad (6.3a)$$

[11] We recall that $\vec{q} = (q_x, q_y, 0) = (\mathbf{q}, 0)$ is a vector *on the standard screen* whose z–component is always zero. We shall use boldface $\mathbf{q}$ to indicate the two-vector $\mathbf{q} = (q_x, q_y)$.

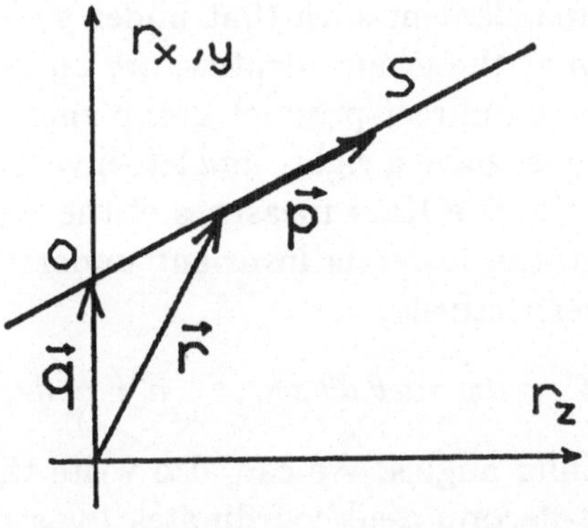

FIGURE 3. The vector $\vec{r}$ on a ray, its length parameter s from the screen, and its position $\vec{q} = (q_x, q_y, 0)$ on the standard screen $r_z = 0$.

$$= \quad (\vec{r} - s'\vec{p}/n)\mathbf{R} + \vec{v} = (\vec{q} + [s - s']\vec{p}/n)\mathbf{R} + \vec{v} \tag{6.3b}$$

$$= \quad \left(\vec{q} + \left[\frac{r_z}{p_z} - \frac{r'_z}{p'_z}\right]\vec{p}\right)\mathbf{R} + \vec{v}, \tag{6.3c}$$

that is independent of s, with the third component zero.

As a particular case we may obtain from (6.3) explicitly the transformation of the rays in $\wp$ under translations along the optical axis by $\mathsf{E}(\mathbf{1}, (0, 0, v_z))$:

$$\vec{p} \mapsto \vec{p}' = \vec{p}, \qquad \mathbf{q} \mapsto \mathbf{q}' = \mathbf{q} - v_z/p_z\, \mathbf{p}, \tag{6.4}$$

where $\mathbf{q} = (q_x, q_y)$ and $\mathbf{p} = (p_x, p_y)$. This is geometrically obvious in Figure 3, but underscores the fact that while $r_z \mapsto q'_z = r_z + v_z$ for any point $\vec{r}$ of the bearing space, the *screen*-coordinate $\mathbf{q}$ of the *ray* slides down; thence the minus sign. If we want to *advance the screen* along the optical axis, we should translate the space by $-z$.

The detailed formulas for the general transformation of the ray and coset coordinates under elements g of the Euclidean group may be found from ordinary vector analysis and are, by themselves, not particularly succint. What is important is that the ray coordinates $\{\vec{p}(\theta, \phi), \vec{q}\}$ map only amongst themselves while the coordinates in the coset $\{\psi, s\}$ transform according to the coset to which they belong, *i.e.*, $\psi'(\psi, \vec{p}(\theta, \phi); g)$ and $s'(s, \vec{p}, \vec{q}; g)$. This is a general consequence of the fact that the spaces of cosets are base spaces for the group bundle. Another consequence of this pertains the factorization of the Haar measure as we shall now show for the $\wp$ manifold of cosets in $\mathcal{E}_3$.

We may ask from (6.1)–(6.3) if there is a volume element in $\wp$, the space of geometric optics rays, that is invariant under the Euclidean group; if so, this will provide a good integration measure for purposes of harmonic analysis and Wigner distribution theory. In fact, such exists and appears to be the origin of the *symplectic* structure of $\wp$. The Haar measure dg of a group manifold is a volume element such that under $g \mapsto g' = gg_0$ or $g \mapsto g'' = g_0^{-1}g$, the Jacobians at the group identity, are unity: $\partial g/\partial g'|_{g=e} = 1$ or $\partial g/\partial g''|_{g=e} = 1$ [2]. A semidirect-product group such as the Euclidean groups in any dimension does have a right- *and* left-invariant Haar measure that is the simple product of the Haar measures of the rotation and translation subgroups, because the latter is invariant under the former. They are, for $\mathcal{E}_3$, $\mathcal{R}_3$, and $\mathcal{T}_3$, respectively,

$$dg = d^3\mathbf{P}\, d^3\vec{r}, \qquad d^3\mathbf{P} = d\psi \sin\theta\, d\theta\, d\phi, \qquad d^3\vec{r} = dr_x\, dr_y\, dr_z. \tag{6.5}$$

where $\{\psi, \theta, \phi\}$ are the Euler angles. We can also write the invariant measure in terms of the coset-decomposed coordinates $\{\psi, s; \vec{p}, \vec{q}\}$. For the rotation subgroup, recalling (3.5*b*),

$$d^3\mathbf{R} = d\psi\, \frac{dp_x\, dp_y}{np_z}. \tag{6.6}$$

The expression for $d^3\vec{r}$ in terms of $ds\, dq_x\, dq_y$ can be found from (5.4),

$$d^3\vec{r} = (dq_x + p_x/n\, ds + s\, dp_x/n) \wedge (dq_y + p_y/n\, ds + s\, dp_y/n) \wedge (p_z\, ds + s\, dp_z)/n, \tag{6.7}$$

where $dp_z = -(p_x\, dp_x + p_y\, dp_y)/p_z$. Using differential form calculus [5], where only unlike-differential factors remain, the outer product of (6.6) and (6.7) is then the Euclidean-invariant Haar measure

$$dg = d\psi\, ds\, n^{-2}\, dp_x\, dp_y\, dq_x\, dq_y. \tag{6.8}$$

The factor $d\psi\, ds$ is the volume element over the space of each coset, while the second factor is the Euclidean-invariant volume element over the space of rays, $\wp$.

6.7 The Euclidean algebra generators on rays

In Section 4 we displayed the generators of the Euclidean algebra on the $\mathcal{E}_3$ manifold $\{\mathbf{P}, \vec{r}\}$. Now that we have found how the rays of geometric optics live on the submanifold $\wp \subset \mathcal{E}_3$, we shall restrict the generators of $\mathcal{E}_3$ to this, by *elliminating* p_z and r_z. We will arrive thus at those previously found for geometric optics [6].[12]

The direction manifold, the Descartes sphere, is a two-dimensional manifold. We have written on occasion $\vec{p}(\theta, \phi)$, but for a large neighborhood $(0 \le \theta < \pi/2)$ of the optical axis $(\theta = 0)$ we can work equivalently with $\mathbf{p} = (p_x, p_y)$, since the third component is $p_z = \sqrt{n^2 - p^2}$ (we denote $p = |\mathbf{p}| \le n$). If we go beyond this *forward hemisphere*, into the

[12] We note a difference of a $\pi/2$ rotation around the z axis with respect to this reference.

backward hemisphere, we must supplement the coordinates in $\wp$ by a *sign* $\sigma \in \{+,0,-\}$, $p_z = \sigma\,|p_z|$, as done in Section 2. All operators on $\wp$ involving $\partial/\partial\mathbf{p}$ are thus in principle *two-chart* operators, with forms perhaps differing by a sign on the two hemispheres of the Descartes sphere, and matching requirements for functions $f(\mathbf{p},\mathbf{q},\sigma)$ near the equator. In this Section we shall not build a Hilbert space of such functions, however, so these precautions are not as indispensable as they will be for Helmholtz optics in Section 9 *et seq.*

On functions $f(p_x,p_y,q_x,q_y;\sigma)$ of $\wp$, $\partial f/\partial p_z$ will be zero. Hence, in subducing operators from $\mathcal{E}_3$ to $\wp$, the restrictions on $\vec{p}$ and $\partial/\partial\vec{p}$ are:

$$p_z \mapsto h = \pm\sqrt{n^2 - p^2}, \qquad \frac{\partial}{\partial p_z} \mapsto 0. \tag{7.1}$$

For the Euclidean position manifold $\vec{r} \in \Re^3$ we similarly reduce the independent position variables $\vec{r} = (r_x, r_y, r_z)$ to the two coordinates on the standard screen $\mathbf{q} = (q_x, q_y)$ and length along the ray s through (5.4), and thereafter require the independence of our function space on s. The action of $\partial/\partial r_z$ on functions f of $\mathbf{q}$ is given by the chain rule for (5.4*a*) and (5.4*b*), namely

$$\frac{\partial f(\mathbf{q})}{\partial r_z} = \frac{\partial \mathbf{q}}{\partial r_z}\cdot\frac{\partial f(\mathbf{q})}{\partial \mathbf{q}} = -\frac{\mathbf{p}}{h}\cdot\frac{\partial f(\mathbf{q})}{\partial \mathbf{q}}. \tag{7.2}$$

The proper replacement of $\vec{r}$ by $\mathbf{q}$ thus entails[13]

$$r_z \mapsto 0, \qquad \frac{\partial}{\partial r_z} \mapsto -\frac{\mathbf{p}}{h}\cdot\frac{\partial}{\partial \mathbf{q}}. \tag{7.3}$$

The generators of the Euclidean translations, Eqs. (4.1*b*) on the *screen* variables of $\wp$, are[14]

$$\hat{T}_x^{\wp} = \frac{\partial}{\partial q_x} = -\{p_x, \circ\}, \tag{7.4a}$$

$$\hat{T}_y^{\wp} = \frac{\partial}{\partial q_y} = -\{p_y, \circ\}, \tag{7.4b}$$

$$\hat{T}_z^{\wp} = -\frac{\mathbf{p}}{h}\cdot\frac{\partial}{\partial \mathbf{q}} = \{h, \circ\}. \tag{7.4c}$$

---

[13]Notice *very* carefully that, as announced, the operator exhibiting h is actually a *two-chart* operator, having two different signs when acting on the forward and the backward hemispheres of the Descartes sphere of ray directions.

[14]The eye catches an apparent *sign* difference between the last expressions in (7.4*a*, *b*) and in (7.4*c*). To set intuition straight, look at Figure 3, displacing space with the embedded rays in x–y and z directions from a fixed reference screen. In the former cases the intersection moves *with* the space, while in the latter it *slides* down along $\mathbf{p}$ with a factor $-|\mathbf{p}|/h = -\tan\theta$.

The generators of rotations are on $\wp$ given by Eqs. (4.3) with the replacements (7.3). They are

$$\hat{R}_x^{\wp} = -h\frac{\partial}{\partial p_y} - q_y\frac{\mathbf{p}}{h}\cdot\frac{\partial}{\partial \mathbf{q}} = \{-q_y h, \circ\}, \tag{7.5a}$$

$$\hat{R}_y^{\wp} = h\frac{\partial}{\partial p_x} + q_x\frac{\mathbf{p}}{h}\cdot\frac{\partial}{\partial \mathbf{q}} = \{q_x h, \circ\}, \tag{7.5b}$$

$$\hat{R}_z^{\wp} = p_x\frac{\partial}{\partial p_y} - p_y\frac{\partial}{\partial p_x} + q_x\frac{\partial}{\partial q_y} - q_y\frac{\partial}{\partial q_x} = \{\mathbf{p}\times\mathbf{q}, \circ\}. \tag{7.5c}$$

The last term in each line of the above six expressions, writes the Euclidean group generators as *Poisson* operators of functions $f(\mathbf{p},\mathbf{q})$, with the symbol

$$\{f, \circ\} = \frac{\partial f}{\partial \mathbf{q}}\cdot\frac{\partial}{\partial \mathbf{p}} - \frac{\partial f}{\partial \mathbf{p}}\cdot\frac{\partial}{\partial \mathbf{q}}. \tag{7.6}$$

That such can be done is in principle quite remarkable, although our acquaintance with Hamiltonian optics makes us expect this to happen, and rightfully call the manifold $(\mathbf{p},\mathbf{q})$ the *phase space* of geometrical optics. What it means is that Euclidean motions, in particular screen motion along the z–axis (6.4), is a *canonical* evolution of the system governed by a Hamiltonian h and conserving the phase space volume element $d\mathbf{p}\,d\mathbf{q}$, as seen at the end of last Section. The Euclidean Lie algebra is thus the natural *dynamical* algebra of the manifold of rays $\wp$ in a homogeneous medium, with z taking the role of time. Indeed, *Hamilton*'s equations hold on $\wp$ with canonically conjugate coordinates $\mathbf{p}$ and $\mathbf{q}$. The first of these equations, on $d\mathbf{q}/dz$, is found from (5.4a,b). Since r_x, r_y and $\vec{p}$ are independent of $r_z = z$, the total and partial derivatives with respect to this z are the same. It follows through (7.1) that

$$\frac{d\mathbf{q}}{dz} = -\frac{\mathbf{p}}{h} = \frac{\partial h}{\partial \mathbf{p}}. \tag{7.7}$$

It is known that the origin of this *first* equation is geometrical [7]. The second, *dynamical*, Hamilton equation is of the form $d\mathbf{p}/dz = -\partial h/\partial\mathbf{q}$. It is trivially satisfied in a homogeneous medium since $\vec{p}$ is constant, so $d\mathbf{p}/dr_z = 0$, and also $\partial h/\partial\mathbf{q}$ is zero.[15]

Being familiar with the Hamiltonian formalism, we know that the invariance of this measure extends to *all* transformations generated by operators $\{f,\circ\}$. This is not sufficient, however, to guarantee that $\mathbf{p}'(\mathbf{p},\mathbf{q}) = \exp\{f,\circ\}\mathbf{p}$, for arbitrary f will remain within a disk of radius n, *i.e.*, that

[15]In *inhomogeneous* media, translating from place to place will change the size of the Descartes sphere, $n(\vec{r})$; the vector $\vec{p}$ will accomodate according to Snell's law, conserving the components of $\vec{p}$ perpendicular to $\vec{\nabla}n$, and leading to Hamilton's second equation in nontrivial form [7].

it will be an *optical* transformation mapping $\wp$ onto itself. That f's beyond the Euclidean algebra are permissible will be seen below, in Sections 12 *et seq.*, regarding the generators of a Lorentz group. The Poisson bracket formalism is rooted in the Heisenberg-Weyl algebra, that is royal road to quantization. It will become clear in Section 11, when we draw the way to wavization provided by the *Euclidean* algebra, that screen coordinates are *not* fit to follow it because they are not operators within this algebra.

As a concrete example, we produce a finite translation by z along the optical axis of the space that bears the rays by exponentiating its infinitesimal generator (7.4c) on the screen coordinates. We have:

$$\mathbf{p} \;\mapsto\; \mathbf{p}' = \exp(z\hat{T}_z^{\wp})\,\mathbf{p} = \sum_{m=0}^{\infty} \frac{z^m}{m!}\left(-\frac{\mathbf{p}}{h}\cdot\frac{\partial}{\partial\mathbf{q}}\right)^m \mathbf{p} = \mathbf{p}, \tag{7.8a}$$

$$\mathbf{q} \;\mapsto\; \mathbf{q}' = \exp(z\hat{T}_z^{\wp})\,\mathbf{q} = \left(1 - z\frac{\mathbf{p}}{h}\cdot\frac{\partial}{\partial\mathbf{q}} + \cdots\right)\mathbf{q} = \mathbf{q} - z\frac{\mathbf{p}}{h}. \tag{7.8b}$$

This result is the same as (6.4), and is geometrically obvious.

6.8 The coset space of wavefront optics

We shall now follow the cosetting strategy seen above to describe other kinds of optics where the elementary objects are not lines, but *planes*, *i.e.*, *wavefronts*. The symmetry group $\mathcal{H}^{\mathrm{wf}}$ of a plane wavefront is the two-dimensional Euclidean group $\mathcal{E}_2$.

Consider the two-dimensional Euclidean subgroup $\mathcal{H}^{\mathrm{wf}} = \mathcal{E}_2 \subset \mathcal{E}_3$ given by the elements $\mathsf{E}(\mathbf{R}_z(\psi), (t_x, t_y, 0))$, $\psi \in \mathcal{S}_1$, $(t_x, t_y) \in \Re^2$ and a generic decomposition of the elements of $\mathsf{E}(\mathbf{P}, \vec{r}) \in \mathcal{E}_3$ that is paralell to (5.2). The symmetry subgroup factor and a representative of the coset space $\mathcal{W} = \mathcal{E}_2\backslash\mathcal{E}_3$, are

$$\mathsf{E}(\mathbf{P}(\psi,\theta,\phi),(r_x,r_y,r_z)) = \mathsf{E}(\mathbf{R}(\psi,0,0),(t_x,t_y,0))\mathsf{E}(\mathbf{P}(0,\theta,\phi),(0,0,u)). \tag{8.1}$$

The factorization of the rotation subgroup into a polarization angle and a coset representative of the two-sphere, $\vec{p}(\theta,\phi) \in \mathcal{R}_2\backslash\mathcal{R}_3 = \mathcal{S}_2$, proceeds as in Section 5.

Regarding the position parameters $\vec{r}$, the analogue of Eqs. (5.4) is

$$(r_x, r_y, r_z) = (t_x, t_y, 0)\mathbf{P}(0,\theta,\phi) + (0,0,u). \tag{8.2a}$$

Directly replacing the matrix $\mathbf{P}(0,\theta,\phi) = \mathbf{R}_y(\theta)\mathbf{R}_z(\phi)$ found through (4.2), we obtain

$$(r_x, r_y) = (t_x\cos\theta, t_y)\begin{pmatrix}\cos\phi & \sin\phi\\ -\sin\phi & \cos\phi\end{pmatrix}, \quad r_z = -t_x\sin\theta + u. \tag{8.2b}$$

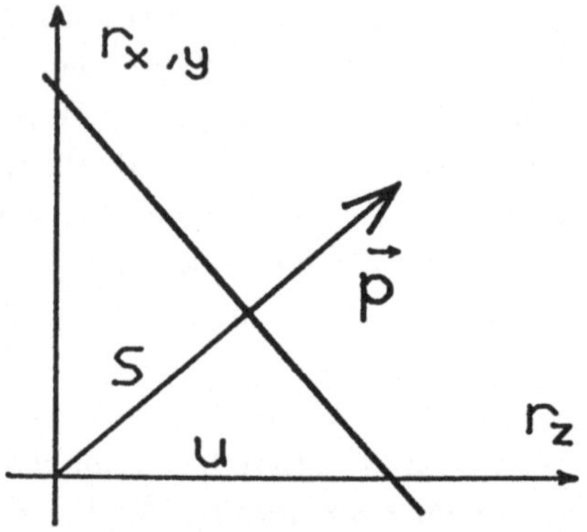

FIGURE 4. Plane wavefronts in space showing the coset space parameters $\vec{p}$ and u, as well as $s = up_z/n$.

These can be inverted for the new parameters in terms of $\mathbf{p} = (p_x, p_y)$, $\mathbf{r} = (r_x, r_y)$, dot and cross products, as

$$t_x = \frac{n}{p_z}\frac{\mathbf{p}}{|\mathbf{p}|}\cdot\mathbf{r}, \qquad t_y = \frac{\mathbf{p}}{|\mathbf{p}|}\times\mathbf{r}, \qquad u = r_z + \frac{\mathbf{p}\cdot\mathbf{r}}{p_z} = \frac{\vec{p}\cdot\vec{r}}{p_z}. \tag{8.2c}$$

The coordinates $\{\psi, t_x, t_y\}$ are in the $\mathcal{H}^{\rm wf}$ coset and $\{\vec{p}(\theta,\phi), u\}$ in the space of cosets $\mathcal{W} = \mathcal{E}_2\backslash\mathcal{E}_3$. The decomposition fails when $p_z = 0$, *i.e.*, for planes parallel to the z–axis. This was also a feature of geometric optics.

The points in $\mathcal{W}$, $\mathcal{H}^{\rm wf}\mathsf{E}(\mathbf{P}(0,\theta,\phi),(0,0,u))$, as in (8.1), are each a *wavefront*, *i.e.*, a plane $\mathcal{H}^{\rm wf}$ in space, given by $\vec{r}\cdot\vec{p} = up_z$, orthogonal to $\vec{p}$, and with intercept u on the z–axis. See Figure 4. The quantity $up_z = \vec{r}\cdot\vec{p} = ns$ is n times the distance s of the plane to the origin, *i.e.*, ns is the usual *optical distance*. We may picture a function $S_{\vec{p}}(s) = f(\vec{p}, u = ns/p_z)$ as representing a *signal* through a train of paralell planes in the direction $\vec{p}$. The coordinates (t_x, t_y) and polarization angle ψ on the plane are absent in $\mathcal{W}$, of course. On the $r_z = 0$ screen, where the r_x and r_y coordinates may be called q_x and q_x as in the last Section, this plane $\{\vec{p}, u\}$ cuts the line $p_x q_x + p_y q_y = \mathbf{p}\cdot\mathbf{q} = up_z$.

The translation subgroup $\mathcal{T}_3 \subset \mathcal{E}_3$ acts in the following way as a Lie transformation through the decomposition (8.1):

$$\mathsf{E}(\mathbf{1},\vec{v})f(\vec{p},u) = f[\mathcal{H}^{\rm wf}\mathsf{E}(\vec{p},(0,0,u))\mathsf{E}(\mathbf{1},\vec{v})] = f(\vec{p}, u + \vec{v}\cdot\vec{p}/p_z). \tag{8.3}$$

The rotations that transform $\vec{p} \mapsto \vec{p}\,' = \vec{p}\mathbf{R}$ and $\vec{r} \mapsto \vec{r}\,' = \vec{r}\mathbf{R}$, will turn $u = \vec{r}\cdot\vec{p}/p_z$ into $u' = \vec{r}\,'\cdot\vec{p}\,'/p_z' = \vec{r}\cdot\vec{p}/p_z' = up_z/p_z'$ and $s = s'$. Hence,

$$\mathsf{E}(\mathbf{R},\vec{0})f(\vec{p},u) = f[\mathcal{H}^{\rm wf}\mathsf{E}(\vec{p},(0,0,u))\mathsf{E}(\mathbf{R},\vec{0})] = f(\vec{p}\mathbf{R}, up_z/p_z'). \tag{8.4}$$

We now realize the Lie algebra of the Euclidean group $\mathcal{E}_3$ on the space of wavefronts $\mathcal{W}$. We may find the Lie generators as in last Section, or *ab initio*, since the group action in (8.3) and (8.4) is explicit. For the translations, as in (4.1) and (7.4),

$$\hat{T}_x^{\mathcal{W}} = \frac{p_x}{p_z}\frac{\partial}{\partial u}, \qquad \hat{T}_y^{\mathcal{W}} = \frac{p_y}{p_z}\frac{\partial}{\partial u}, \qquad \hat{T}_z^{\mathcal{W}} = \frac{\partial}{\partial u}. \tag{8.5a, b, c}$$

For the rotations (8.4) we can compute explicitly the scale factor p_z/p'_z for $\mathbf{R}_x(\alpha_x)$, $\mathbf{R}_y(\alpha_y)$, and $\mathbf{R}_z(\alpha_z)$ from the matrices (4.2) acting on the row vector $\vec{p}$, and then let $\alpha \to 0$. The results are:

$$\hat{R}_x^{\mathcal{W}} = p_y\frac{\partial}{\partial p_z} - p_z\frac{\partial}{\partial p_y} - \frac{p_y}{p_z}u\frac{\partial}{\partial u}, \tag{8.6a}$$

$$\hat{R}_y^{\mathcal{W}} = p_z\frac{\partial}{\partial p_x} - p_x\frac{\partial}{\partial p_z} + \frac{p_x}{p_z}u\frac{\partial}{\partial u}, \tag{8.6b}$$

$$\hat{R}_z^{\mathcal{W}} = p_x\frac{\partial}{\partial p_y} - p_y\frac{\partial}{\partial p_x}. \tag{8.6c}$$

We may check that the Lie brackets (4.4) hold.

6.9 Helmholtz optics

The Euclidean generators (8.5)–(8.6) form a Lie algebra, and within its *enveloping* algebra we can find group *invariants*. The three-dimensional Euclidean algebra has two quadratic invariants, $\hat{T}^2 = \vec{\hat{T}}\cdot\vec{\hat{T}}$ and $\vec{\hat{T}}\cdot\vec{\hat{R}}$, that may be used to label the irreducible representations of the algebra and group. The latter invariant is identically zero on $\mathcal{W}$. The former, on the other hand, is

$$\hat{T}^2 = (\hat{T}_x^{\mathcal{W}})^2 + (\hat{T}_y^{\mathcal{W}})^2 + (\hat{T}_z^{\mathcal{W}})^2 = \frac{n^2}{p_z^2}\frac{\partial^2}{\partial u^2}. \tag{9.1}$$

The action of the Euclidean group on $\mathcal{W}$ will thus map functions $f(\vec{p}, u)$ amongst themselves, but respecting the linear *eigenspaces* of the operator in (9.1), that will remain invariant under that process. Their u-dependent factors will be linear combinations of $\sin kp_z u/n$ and $\cos kp_z u/n$, for k in principle complex, with eigenvalues $-k^2$. We may thus apply Fourier analysis in $u \in \Re$ to the functions f on $\mathcal{W}$. Each of its partial wave components $\mathcal{W}_k$, is the form

$$f(\vec{p}, u) = \Phi(\vec{p})\exp iks, \qquad s = up_z/n = \vec{p}\cdot\vec{r}/n, \tag{9.2}$$

and they will transform irreducibly under the Euclidean group. The Euclidean invariant k is the *wavenumber* in the medium of refractive index n. When this number k is real, the functions (9.2) exhibit a *translational invariance* under $s \mapsto s + \lambda$, $\lambda = 2\pi/k$, as true plane *waves* do.[16] We thus have all and only plane waves of a given wavenumber, in all directions of the Descartes sphere.

In the same way as in geometric optics, the direction sphere $\vec{p} \in \mathcal{S}_2$ may be projected (twice) on its equatorial screen plane, the disk δ_n, where $|\mathbf{p}| < n$ and the boundary $|\mathbf{p}| = n$ sews the two disks. We will write the function $\Phi(\vec{p})$ as $\Phi_\pm(\mathbf{p})$ independent of p_z through for $p_z \mapsto \sigma\sqrt{n^2 - p^2}$ (with $p^2 = \mathbf{p}\cdot\mathbf{p}$); the *sign* of p_z, $\sigma \in \{+, 0, -\}$ distinguishes the two open hemispheres and the common boundary circle. (We shall usually disregard the zero.) The functions will be understood to be continuous between the two charts, *i.e.*, matching as $\lim_{|\mathbf{p}|\to n} \Phi_+(\mathbf{p}) = \lim_{|\mathbf{p}|\to n} \Phi_-(\mathbf{p}) = \Phi_0(\mathbf{p})$.

As in ordinary Fourier analysis, the operator $\partial/\partial u$ acts only on the e^{iks} factor (recall that $s = up_z/n$), so in the translation generators (8.5), it is replaced by the factor ikp_z/n, with one sign in each chart. The generators of Euclidean translations in $\mathcal{W}_k$ are thus

$$\hat{T}_x^{k\pm} = \frac{ikp_x}{n}, \qquad \hat{T}_y^{k\pm} = \frac{ikp_y}{n}, \qquad \hat{T}_z^{k\pm} = \frac{\pm ik\sqrt{n^2 - p^2}}{n}. \tag{9.3}$$

The restriction from $\mathcal{W}$ to $\mathcal{W}_k$ of the generators of rotation (8.6) proceeds through noting that neither $\partial/\partial p_z$ nor $\partial/\partial u$ act on $\Phi_\pm(\mathbf{p})$, and that $p_z\partial s/\partial p_z = u\partial s/\partial u$. Hence, their form on functions of this space is:

$$\hat{R}_x^{k\pm} = \mp\sqrt{n^2 - p^2}\,\frac{\partial}{\partial p_y}, \tag{9.4a}$$

$$\hat{R}_y^{k\pm} = \pm\sqrt{n^2 - p^2}\,\frac{\partial}{\partial p_x}, \tag{9.4b}$$

$$\hat{R}_z^{k\pm} = p_x\frac{\partial}{\partial p_y} - p_y\frac{\partial}{\partial p_x}. \tag{9.4c}$$

This realization of the Euclidean algebra, by construction, belongs to a definite irreducible representation, determined by the values of the Casimir invariants. Plane waves moreover, are a representation *basis* further reduced and classified by the translation subalgebra $\hat{T}_j^{k\pm}$, $j = x, y$, and the sign of $\hat{T}_z^{k\pm}$. Another subgroup basis, where the diagonal generators are $\hat{T}_z^{k\pm}$ and $\hat{R}_z^{k\pm}$, are functions with support on a ring $\theta = \theta_0$, with a definite rotation covariance. These are the *nondiffracting J_m-beams* [8]. Finally, *multipole fields* are obtained as the rotation subgroup eigenbasis of the subalgebra

[16] We note that there is a fundamental wavenumber (light *color*) k_0 associated with the vacuum $n = 1$, and for any other medium $k = n\,k_0$. Thus even though k and n will be written jointly in most of the text, they always appear as the ratio $k/n = k_0$.

Casimir operator

$$\sum_{j=x,y,z} (\hat{R}_j^{k\pm})^2 = n^2 \left(\frac{\partial^2}{\partial p_x^2} + \frac{\partial^2}{\partial p_x^2} \right) - \left(p_x \frac{\partial}{\partial p_x} + p_y \frac{\partial}{\partial p_y} \right) \left(p_x \frac{\partial}{\partial p_x} + p_y \frac{\partial}{\partial p_y} + 1 \right) \quad (9.5)$$

and $\hat{R}_z^{k\pm}$, *i.e.*, the projection of the spherical harmonics of the sphere on the disk. Each of these bases will yield a different realization of the Euclidean algebra through differential or difference operators in the row labels of the representation.

The effect of the *exponentials* of these operators on the two-chart functions are simple when the direction hemispheres do not mix, and rather complicated otherwise. In any case, we may conveniently revert to the description by functions $\Phi(\vec{p})$ over the sphere. Translations $\exp(\sum_j v_j \hat{T}_j)$ multiply it with the phase $\exp(ik \sum_j v_j p_j / n) = \exp(ik_0 \sum_j v_j p_j)$, and rotations act on the argument row vector $\vec{p}$ through right matrix multiplication as usual.

6.10 The Hilbert space for Helmholtz optics

Quantum mechanics works with $\mathcal{L}^2(\Re^n)$ Hilbert spaces of wavefunctions where real observables are eigenvalues of self-adjoint operators, and symmetry transformations are unitary. We shall now proceed to build a Hilbert space for the oscillatory solutions of the Helmholtz equation that is unitarily equivalent to $\mathcal{L}^2(\mathcal{S}_2)$, the well-known space of square-integrable functions on the Descartes direction sphere. We call the structure *Helmholtz optics.*

Let us return to the Haar measure over $\mathcal{E}_3$ given in (6.5)–(6.6). On the direction sphere $|\vec{p}| = n$, the invariant *surface* element is

$$d^2 S(\vec{p}) = n^2 \sin\theta \, d\theta \, d\phi = \frac{n}{p_z} \, dp_x \, dp_y. \quad (10.1)$$

In the second form, the Descartes sphere surface element has been projected over the screen plane as before. We must specify that $p_z > 0$ in the 'forward' hemisphere $0 \leq \theta < \pi/2$, and $p_z < 0$ in the 'backward' hemisphere $\pi/2 < \theta \leq \pi$, taking account of the change in the surface element orientation.

A continuous linear superposition $\Phi(\vec{p})$ of plane waves (9.2) over all directions is in $\Re^3$,[17]

$$F(\vec{r}) = \frac{k}{2\pi n} \int_{\mathcal{S}_2} d^2 S(\vec{p}) \Phi(\vec{p}) \exp(ik\vec{p} \cdot \vec{r}/n), \quad (10.2)$$

[17] We choose the normalization factor for the purposes of symmetry in the Fourier analysis formulas. Position $\vec{r}$ has units of $k^{-1} = \lambda/2\pi$, the reduced wavelength. We may ascribe to $\vec{p}$ units of n, although physically dimensionless. Inte-

and satisfies the Helmholtz equation in this space:

$$\left(\frac{\partial^2}{\partial r_x^2}+\frac{\partial^2}{\partial r_y^2}+\frac{\partial^2}{\partial r_z^2}\right)F(\vec{r})=-k^2F(\vec{r}). \qquad (10.3)$$

This may be also written in *evolution* form on a space of two-component functions

$$\begin{pmatrix}0 & 1\\ -\Delta_k & 0\end{pmatrix}\begin{pmatrix}F(\vec{r})\\ F'(\vec{r})\end{pmatrix}=\frac{\partial}{\partial r_z}\begin{pmatrix}F(\vec{r})\\ F'(\vec{r})\end{pmatrix},\quad \Delta_k=k^2+\frac{\partial^2}{\partial r_x^2}+\frac{\partial^2}{\partial r_x^2}. \qquad (10.4)$$

The first component of this equation defines $F'(\vec{r})$ to be the r_z–derivative of $F(\vec{r})$, while the second reproduces the Helmholtz equation (10.3).

Previously, the geometric optics model reduced the description of rays in position 3-space to a standard 2-dimensional *screen* at $r_z=0$. This suggests that we perform the Helmholtz analogue of this *regression to screen values* $\vec{r}\mapsto(\mathbf{q},0)$ —with some extra care due to the *two-chart* structure pointed out before. Oscillating solutions to the Helmholtz equation are determined thoughout $\Re^3$ by specifying their initial value and normal derivative at a plane. From (10.1) and (10.2) these are, *on the screen* $r_z=0$ and expressed as integrals over the $\mathbf{p}$–*disk* δ_n of radius n,[18]

$$\begin{aligned}F(\mathbf{q}) &= F(\vec{r})\big|_{r_z=0}\\ &= \frac{k}{2\pi}\int_{\delta_n}\frac{d^2\mathbf{p}}{\sqrt{n^2-p^2}}\left[\Phi_+(\mathbf{p})+\Phi_-(\mathbf{p})\right]e^{ik\mathbf{p}\cdot\mathbf{q}/n}, \qquad (10.5a)\\ F'(\mathbf{q}) &= \frac{\partial F(\vec{r})}{\partial r_z}\bigg|_{r_z=0}\\ &= \frac{ik^2}{2\pi n}\int_{\delta_n}d^2\mathbf{p}\,\left[\Phi_+(\mathbf{p})-\Phi_-(\mathbf{p})\right]e^{ik\mathbf{p}\cdot\mathbf{q}/n}. \qquad (10.5b)\end{aligned}$$

Both the function $F(\mathbf{q})$ at the screen *and* its normal derivative $F'(\mathbf{q})$ are needed to encode the information contained in the two functions $\Phi_\pm(\mathbf{p})$ on the disk δ_n. The inversion of (10.5) to solve for the function on the sphere is found through Fourier transformation in the plane:

$$\Phi_\pm(\mathbf{p})=\frac{k}{4\pi n}\int_{\Re^2}d^2\mathbf{q}\left[\frac{\sqrt{n^2-p^2}}{n}F(\mathbf{q})\pm\frac{1}{ik}F'(\mathbf{q})\right]e^{-ik\mathbf{p}\cdot\mathbf{q}/n}. \qquad (10.6)$$

---

gration over the sphere with the measure (10.1) endows it with units of n^2. Hence, if we regard $\Phi(\vec{p})$ as having units n^{-1}, $F(\vec{r})$ will have units of k. We recall that $k/n=k_0$, the fixed wavenumber of vacuum in the Fourier signal decomposition of last Section.

[18]Note the factor $k/2\pi n=k_0/2\pi=1/\lambda_0=1/n\lambda$, with λ_0 the wavelength in vacuum $n=1$.

For example, a single plane wave $\Omega_{\vec{p}_0}$ directed by some $\vec{p}_0$ will have its coset representative function given by a Dirac delta on the Descartes sphere, *i.e.*, under the measure (10.1) and appropriate range,

$$\Omega_{\vec{p}_0}(\vec{p}) = \delta_{S_2}(\vec{p}_0, \vec{p}) = (n^2 \sin\theta)^{-1}\delta(\theta_0 - \theta)\delta(\phi_0 - \phi), \quad (10.7a)$$

or,

$$\Omega_{\mathbf{p}_0,\sigma}(\mathbf{p}) = \frac{\sqrt{n^2 - p^2}}{n}\,\delta(p_{0x} - p_x)\,\delta(p_{0y} - p_y)\delta_{\sigma,\mathrm{sign}\,p_{0z}}. \quad (10.7b)$$

The corresponding Helmholtz plane-wave solution function and its normal derivative at the screen form then the two-function

$$\mathbf{W}_{\vec{p}_0}(\mathbf{q}) = \begin{pmatrix} W_{\vec{p}_0}(\mathbf{q}) \\ W'_{\vec{p}_0}(\mathbf{q}) \end{pmatrix} = \frac{k}{2\pi n}\begin{pmatrix} 1 \\ ikp_{0z}/n \end{pmatrix} e^{ik\mathbf{p}_0\cdot\mathbf{q}/n}. \quad (10.8)$$

The normal derivative distinguishes the two distinct plane waves $\vec{p}^+ = (\mathbf{p}, p_z)$ and $\vec{p}^- = (\mathbf{p}, -p_z)$ that are *reflected* versions of each other by a mirror in the screen. A Helmholtz function whose normal derivative on the screen is *zero* contains, for every constituent plane wave, its reflection. The last Section will elaborate further on this.

Square-integrable functions on the sphere with the measure (10.1) are well known to constitute a Hilbert space $\mathcal{L}^2(\mathcal{S}_2)$, for which a definite value of k is implied. The sesquilinear inner product of two functions in that space is

$$(\Phi_1, \Phi_2)_{S_2} = \int_{S_2} d^2S(\vec{p})\,\Phi_1(\vec{p})^*\,\Phi_2(\vec{p}) \quad (10.9a)$$

$$= \int_{\delta_n} \frac{n\,d^2\mathbf{p}}{\sqrt{n^2 - p^2}} \times [\Phi_{1,+}(\mathbf{p})^*\,\Phi_{2,+}(\mathbf{p}) + \Phi_{1,-}(\mathbf{p})^*\,\Phi_{2,-}(\mathbf{p})], \quad (10.9b)$$

where the asterisk * indicates complex conjugation. This inner product is manifestly invariant under rotations of the Descartes sphere, as well as under translations, since the latter only multiply the functions by a phase that cancels on account of the sesquilinearity of the inner product. It is an invariant under Euclidean transformations: $(\mathcal{L}_g\Phi_1, \mathcal{L}_g\Phi_2)_{S_2} = (\Phi_1, \Phi_2)_{S_2}$. The last expression shows the form of the inner product as an integral over the disk δ_n of radius n, for both 'forward' and 'backward' waves. The Euclidean generators in (9.3) and (9.4) will be skew-adjoint under this inner product.

Let us now write the inner product (10.9) in terms of the initial value and normal derivative on the screen of solutions to the Helmholtz equation, as given by equations (10.5), replacing the $\Phi_\pm(\mathbf{p})$'s from (10.6) into (10.9).

There is a triple integration where we can move to the right the integral over the compact domain,

$$(\Phi_1, \Phi_2)_{S_2} = \left(\frac{k}{2\pi n}\right)^2 \int_{\delta_n} \frac{n\, d^2\mathbf{p}}{\sqrt{n^2 - p^2}} \int_{\Re^2} d^2\mathbf{q} \int_{\Re^2} d^2\mathbf{q}'\, e^{-ik\mathbf{p}\cdot(\mathbf{q}-\mathbf{q}')/n} \times \left[\frac{n^2 - p^2}{n^2} F_1(\mathbf{q})^* F_2(\mathbf{q}') + \frac{1}{k^2} F_1'(\mathbf{q})^* F_2'(\mathbf{q}')\right] \tag{10.10a}$$

$$= \left(\frac{k}{2\pi n}\right)^2 \int_{\Re^2} d^2\mathbf{q} \int_{\Re^2} d^2\mathbf{q}' \times [\omega(|\mathbf{q}-\mathbf{q}'|) F_1(\mathbf{q})^* F_2(\mathbf{q}') + \varpi(|\mathbf{q}-\mathbf{q}'|) F_1'(\mathbf{q})^* F_2'(\mathbf{q}')]\,. \tag{10.10b}$$

We have assimilated the **p**–integration[19] into two *nonlocal weight functions*, ω and ϖ,

$$\begin{aligned}\omega(|\mathbf{q}-\mathbf{q}'|) &= \frac{1}{2}\int_{\delta_n} d^2\mathbf{p}\, \frac{\sqrt{n^2-p^2}}{n} e^{-ik\mathbf{p}\cdot(\mathbf{q}-\mathbf{q}')/n} \\ &= \frac{1}{2}\int_0^n p\, dp \frac{\sqrt{n^2-p^2}}{n} \int_0^{2\pi} d\varphi\, e^{-ikp|\mathbf{q}-\mathbf{q}'|\cos\varphi/n} \\ &= \frac{\pi}{n}\int_0^n p\, dp \sqrt{n^2-p^2}\, J_0(kp|\mathbf{q}-\mathbf{q}'|/n) \qquad (10.11a) \\ &= \pi n^2 \frac{j_1(k|\mathbf{q}-\mathbf{q}'|)}{k|\mathbf{q}-\mathbf{q}'|},\end{aligned}$$

and

$$\begin{aligned}\varpi(|\mathbf{q}-\mathbf{q}'|) &= \frac{1}{2k^2}\int_{\delta_n} d^2\mathbf{p}\, \frac{n}{\sqrt{n^2-p^2}} e^{-ik\mathbf{p}\cdot(\mathbf{q}-\mathbf{q}')/n} \\ &= \frac{\pi n^2}{k^2} j_0(k|\mathbf{q}-\mathbf{q}'|), \qquad (10.11b)\end{aligned}$$

---

[19] We note the useful integral [Gradshteyn & Ryzhik, Eqs. 6.567.1 and 6.554.2]:

$$\int_0^n p\, dp\, (n^2 - p^2)^\mu J_0(xp/n) = 2^\mu \Gamma(\mu+1) n^{2(\mu+1)} \frac{J_{\mu+1}(x)}{x^{\mu+1}}.$$

where we have the spherical Bessel functions

$$j_0(z) = \sqrt{\frac{\pi}{2z}}\,J_{1/2}(z) = \frac{\sin z}{z}, \tag{10.12a}$$

$$\frac{j_1(z)}{z} = \sqrt{\frac{\pi}{2z^3}}\,J_{3/2}(z) = \frac{\sin z - z\cos z}{z^3}. \tag{10.12b}$$

The weight functions are solutions of the Helmholtz equation; $\varpi(\mathbf{q})$ integrates $\Phi(\vec{p}) =$ constant over the whole direction sphere [*cf.* (10.5), thus with zero normal derivative] and $\omega(\mathbf{q})$ correspondingly integrates $\Phi(\vec{p}) =$ constant $\times\, p_z^2$. Since they are the widest, smoothest functions on the sphere, they may be seen as the ***narrowest*** functions on the screen that are still purely oscillatory solutions of the Helmholtz equation.[20]

We may write this inner product on the space of screen conditions for the Helmholtz equation in 2–matrix form as

$$(\mathbf{F}_1,\mathbf{F}_2)_{\mathcal{H}_k} = \int_{\Re^2} d^2\mathbf{q} \int_{\Re^2} d^2\mathbf{q}'\; \mathbf{F}_1(\mathbf{q})^\dagger \mathbf{H}_k(|\mathbf{q}-\mathbf{q}'|)\mathbf{F}_2(\mathbf{q}'), \tag{10.13a}$$

$$\mathbf{F}_j(\mathbf{q}) = \begin{pmatrix} F_j(\mathbf{q}) \\ F_j'(\mathbf{q}) \end{pmatrix}, \tag{10.13b}$$

$$\mathbf{H}_k(|\mathbf{q}-\mathbf{q}'|) = \frac{1}{4\pi}\begin{pmatrix} k^2\dfrac{j_1(k|\mathbf{q}-\mathbf{q}'|)}{k|\mathbf{q}-\mathbf{q}'|} & 0 \\ 0 & j_0(k|\mathbf{q}-\mathbf{q}'|) \end{pmatrix}. \tag{10.13c}$$

This inner product is also *Euclidean invariant*: if $F_j(\vec{r})$, $j = 1,2$ are two *solutions* of the Helmholtz equation, whose values and normal derivatives at the standard screen $r_z = 0$ are $F_j(\mathbf{q})$ and $F_j'(\mathbf{q})$, their inner product is unchanged if we move or rotate the screen to any other plane. Since it was built unitarily equivalent to $\mathcal{L}^2(\mathcal{S}_2)$, it thus serves to define a *Hilbert space* of oscillatory solutions to the Helmholtz equation that we shall call $\mathcal{H}_k$. Such an inner product for the *two*-dimensional Helmholtz solutions was found by Steinberg and Wolf [9] searching for Euclidean-invariant inner products with an in general nonlocal matrix measure $\mathbf{H}_k(|\mathbf{q}-\mathbf{q}'|)$; its matrix elements were obtained through boundary and differential conditions that hold in the subspace of oscillatory solutions of that equation, and shown to be *unique.*[21]

---

[20] Dirac δ's are not allowed in $\mathbf{q}$ since their Fourier conjugate has support outside the $\mathbf{p}$-disk δ_n. Also, evanescent waves are not allowed unless we go into the complex–k extension of our group. This we shall not do here. The issue of ***localizability*** is correspondingly different from what we are familiar with in quantum mechanics.

[21] A similar treatment was made in [9] for the inner product in the Klein-Gordon equation solution space. In the form (10.13), the measure is then shown to be local (by Dirac δ's) and, in matrix form, antidiagonal. This verifies the known result for the three-dimensional Poincaré-invariant inner product. We should note that the inner product is not *total illumination* —that will be examined in the next Section.

6.11 The Euclidean algebra generators in Helmholtz optics —wavization

Our last realization of the Euclidean algebra generators was given in equations (9.3)–(9.4) for two-chart functions on the two disks of the squashed Descartes sphere. We now want to display the form of the generators on the Helmholtz Hilbert space of two-functions on the *screen* $\mathbf{q}$, with the nonlocal inner product (10.13). Upon comparison with the Euclidean generators on the geometric optics phase space, we will arrive at what appears to be a good recipe for wavization.

When the function over the direction sphere $\Phi(\vec{p})$ in (10.2) or its equivalents $\Phi_\pm(\mathbf{p})$ in (10.5), are multiplied by p_x/n or p_y/n, this factor becomes $ik\partial/\partial q_x$ and $ik\partial/\partial q_y$ on the Helmholtz solution $F(\vec{r})$ in (10.2) or its equivalent two-function $\mathbf{F}(\mathbf{q})$ in (10.5). The z-translation generator, multiplication by p_z/n, is different on the two charts: $p_z\Phi(\vec{p}) \mapsto \pm\sqrt{n^2-p^2}\Phi_\pm(\mathbf{p})$. Such a multiplication turns the integral for $F(\mathbf{q})$ in (10.5a) into the integral for $F'(\mathbf{q})$ in (10.5b). It also turns the latter into the former with an integrand factor of $n^2 - p^2$. This factor, in company with $\exp(ik\mathbf{p}\cdot\mathbf{q}/n)$, becomes (minus) the Helmholtz operator Δ_k in equation (10.4) acting on the same exponential. This operator is then extracted from the integral. Hence, the translation generators in (9.3) may be written as 2×2 *matrix* operators on the Helmholtz Hilbert space, thus:

$$\hat{T}_x^{\mathcal{H}_k} = \begin{pmatrix} \partial_{q_x} & 0 \\ 0 & \partial_{q_x} \end{pmatrix}, \quad \hat{T}_y^{\mathcal{H}_k} = \begin{pmatrix} \partial_{q_y} & 0 \\ 0 & \partial_{q_y} \end{pmatrix}, \quad \hat{T}_z^{\mathcal{H}_k} = \begin{pmatrix} 0 & 1 \\ -\Delta_k & 0 \end{pmatrix}. \tag{11.1a, b, c}$$

We may follow a similar procedure for the generators of rotations around the x and y axes in (9.4) through (10.5a,b), the sign difference on the two charts turning one into the other. The roots of $n^2 - p^2$ cancel or combine with their measures; derivatives with respect to p_j's can be integrated by parts because the boundary terms between functions on the two disks cancel, and are thus thrown on the exponential factor. Finally, the exponent turns p_j's into $\partial/\partial q_j$'s and $\partial/\partial p_j$'s into q_j's that can be extracted from the integral. We thus arrive at the following matrix operator realization:

$$\hat{R}_x^{\mathcal{H}_k} = \begin{pmatrix} 0 & q_y \\ -q_y\Delta_k - \partial_{q_y} & 0 \end{pmatrix}, \tag{11.2a}$$

$$\hat{R}_y^{\mathcal{H}_k} = \begin{pmatrix} 0 & -q_x \\ q_x\Delta_k + \partial_{q_x} & 0 \end{pmatrix}, \tag{11.2b}$$

$$\hat{R}_z^{\mathcal{H}_k} = \begin{pmatrix} q_x\partial_{q_y} - q_y\partial_{q_x} & 0 \\ 0 & q_x\partial_{q_y} - q_y\partial_{q_x} \end{pmatrix}. \tag{11.2c}$$

The above generators were also found by Steinberg and Wolf [9], for the two-dimensional Helmholtz equation and were further studied and applied by Atakishiyev, Lassner and Wolf in reference [17]. They are skew-adjoint under the nonlocal inner product in the Helmholtz Hilbert space (10.13). Their commutation relations are of course the same as (4.4), and as a *irreducible representation* of the Euclidean algebra they are identified by their invariants $(\hat{\vec{T}}^{\mathcal{H}_k})^2 = -k^2\mathbf{1}$ and $\hat{\vec{T}}^{\mathcal{H}_k} \cdot \hat{\vec{R}}^{\mathcal{H}_k} = 0$. Regarding the pure rotation subalgebra (11.2), the representation that is spanned is not irreducible, but quite closely so:

$$(\hat{R}_x^{\mathcal{H}_k})^2 + (\hat{R}_y^{\mathcal{H}_k})^2 + (\hat{R}_z^{\mathcal{H}_k})^2 = \begin{pmatrix} \hat{D}(\hat{D}-1) + k^2q^2 & 0 \\ 0 & \hat{D}(\hat{D}+1) + k^2q^2 \end{pmatrix}, \tag{11.3a}$$

where

$$\hat{D} = \tfrac{1}{2}(\mathbf{q}\cdot\partial_{\mathbf{q}} + \partial_{\mathbf{q}}\cdot\mathbf{q}) = q_x\partial_{q_x} + q_y\partial_{q_y} + 1 \tag{11.3b}$$

is a 'dilatation' operator on the screen, that is self-adjoint on $\mathcal{L}^2(\Re^2)$, but not separately so in $\mathcal{H}_k$.

Let us exemplify the handling of the sphere and Helmholtz inner products $(\Phi_1, \Phi_2)_{\mathcal{S}_2} = (\mathbf{F}_1, \mathbf{F}_2)_{\mathcal{H}_k}$ by finding the matrix elements of the z–translation generator in $\mathcal{L}^2(\mathcal{S}_2)$, $\hat{T}_z^{k\pm}$ in Eqs. (9.3), and its Helmholtz version $\hat{T}_z^{\mathcal{H}_k}$ in Eqs. (11.1) through (10.2)–(10.5). This generator is the analogue of the quantum mechanical Hamiltonian, so we may asign $(\Phi, \hat{T}_z^k\Phi)_{\mathcal{S}_2} = (\mathbf{F}, \hat{T}_z^{\mathcal{H}_k}\mathbf{F})_{\mathcal{H}_k}$ the interpretation of the energy —illumination— of the state described by $\Phi(\vec{p})$ as a function of direction, or by $\mathbf{F}(\mathbf{q})$ as the Helmholtz two-function on the screen.

To this end we calculate, following (10.9), the cross matrix elements

$$(\Phi_1, \hat{T}_z^k\Phi_2)_{\mathcal{S}_2} = \int_{\mathcal{S}_2} d^2S(\vec{p})\, \Phi_1(\vec{p})^* \,\frac{ikp_z}{n}\Phi_2(\vec{p}) \tag{11.4a}$$

$$= ik\int_{\delta_n} d^2\mathbf{p}\, [\Phi_{1,+}(\mathbf{p})^*\, \Phi_{2,+}(\mathbf{p}) - \Phi_{1,-}(\mathbf{p})^*\, \Phi_{2,-}(\mathbf{p})]\,. \tag{11.4b}$$

Next, we replace the $\Phi(\mathbf{p})$'s by $F(\mathbf{q})$'s and $F'(\mathbf{q})$'s through (10.6) with cancellation of summands, and exchange integrals. We thus obtain

$$\frac{2k^2}{(4\pi n)^2}\int_{\Re^2} d^2\mathbf{q} \int_{\Re^2} d^2\mathbf{q}' [F_1(\mathbf{q})^* F_2'(\mathbf{q}') - F_1(\mathbf{q})'^* F_2(\mathbf{q}')] \times \int_{\delta_n} d^2\mathbf{p}\, \frac{\sqrt{n^2-p^2}}{n} e^{-ik\mathbf{p}\cdot(\mathbf{q}-\mathbf{q}')/n} \tag{11.4c}$$

$$= \left(\frac{k}{2\pi n}\right)^2 \int_{\Re^2} d^2\mathbf{q} \int_{\Re^2} d^2\mathbf{q}' \times (F_1(\mathbf{q})\; F_1'(\mathbf{q}))^* \begin{pmatrix} 0 & \omega(|\mathbf{q}-\mathbf{q}'|) \\ -\omega(|\mathbf{q}-\mathbf{q}'|) & 0 \end{pmatrix} \begin{pmatrix} F_2(\mathbf{q}') \\ F_2'(\mathbf{q}') \end{pmatrix} \tag{11.4d}$$

$$= \left(\frac{k}{2\pi n}\right)^2 \int_{\Re^2} d^2\mathbf{q} \int_{\Re^2} d^2\mathbf{q}' \, (F_1(\mathbf{q}) \, F_1'(\mathbf{q}))^*$$
$$\times \begin{pmatrix} \omega(|\mathbf{q}-\mathbf{q}'|) & 0 \\ 0 & \varpi(|\mathbf{q}-\mathbf{q}'|) \end{pmatrix} \begin{pmatrix} 0 & 1 \\ -\Delta_k & 0 \end{pmatrix} \begin{pmatrix} F_2(\mathbf{q}') \\ F_2'(\mathbf{q}') \end{pmatrix}. \tag{11.4e}$$
$$= (\mathbf{F}_1, \hat{T}_z^{\mathcal{H}_k} \mathbf{F}_2)_{\mathcal{H}_k}. \tag{11.4f}$$

The step (11.4*c*–*d*) recognizes the integral in (10.11*a*) while the equality (11.4*d*–*e*) proceeds through integration by parts on $\mathbf{q}'$, matrix multiplication, and the differential equality

$$\Delta_k \, \varpi = \omega \tag{11.5}$$

between the two weight functions of the Helmholtz Hilbert space measure.

In the form (11.4*a*) it is evident that we are integrating functions over the sphere of directions with an obliquity factor of $p_z/n = \cos\theta$, where θ is the angle between the plane of the waves and the z = constant screen. Upon transformation to Helmholtz 'wavefunctions' over the screen in (11.4*d*), the same inner product takes a nonlocal Klein-Gordon-type of antidiagonal measure structure [9] that should merit further inquiry elsewhere.

Now that we have presented geometric and Helmholtz optics as two structures contained within the Euclidean group, let us define *wavization* "$\overset{\mathrm{W}}{\mapsto}$" heuristically as the passage from one to the other that is paralell to that from classical to quantum mechanics. Comparison of the translation generators on $\mathcal{W}_k$ given by (9.3) and recall of the relation $k/n = k_0$, suggest the replacement of the 'geometric' momentum components of $\vec{p}$ in the $\wp$ manifold by the self-adjoint matrix operators (11.1) in the Helmholtz Hilbert space $\mathcal{H}_k$, through the map

$$p_x \overset{\mathrm{W}}{\mapsto} -\frac{i}{k_0} \begin{pmatrix} \partial_{q_x} & 0 \\ 0 & \partial_{q_x} \end{pmatrix}, \tag{11.6a}$$
$$p_y \overset{\mathrm{W}}{\mapsto} -\frac{i}{k_0} \begin{pmatrix} \partial_{q_y} & 0 \\ 0 & \partial_{q_y} \end{pmatrix}, \tag{11.6b}$$
$$p_z \overset{\mathrm{W}}{\mapsto} -\frac{i}{k_0} \begin{pmatrix} 0 & 1 \\ -\Delta_k & 0 \end{pmatrix}. \tag{11.6c}$$

Note that the role of Planck's constant $\hbar$ in quantum mechanics is taken by vacuum wavenumber k_0 in Helmholtz screen optics.

Let us articulate first a naïve guess based on our Schrödinger "$\overset{\mathrm{Q}}{\mapsto}$" experience: classical functions satisfying certain Poisson brackets should quantize to operators satisfying analogous commutators. This will turn out to be somewhat off the mark, but may be instructive to point out pitfalls. If we map the space coordinates (r_x, r_y, r_z) to the multiplicative operators

$(\hat{q}_x, \hat{q}_y, 0)$ on functions in $\mathcal{H}_k$, and the symmetrization scheme is followed,[22] the classical components of angular momentum $\vec{R}^\times = \vec{r}\times\vec{p}$ that generate rotations, will map correctly as

$$R_x^\times = r_y p_z - r_z p_y \overset{Q}{\mapsto} \tfrac{1}{2}\{\hat{q}_y, \hat{p}_z\}_+ = -i/k_0\, \hat{R}_x^{\mathcal{H}_k}, \qquad (11.7a)$$
$$R_y^\times = r_z p_x - r_x p_z \overset{Q}{\mapsto} -\tfrac{1}{2}\{\hat{q}_x, \hat{p}_z\}_+ = -i/k_0\, \hat{R}_y^{\mathcal{H}_k}, \qquad (11.7b)$$
$$R_z^\times = r_x p_y - r_y p_x \overset{Q}{\mapsto} \hat{q}_x\hat{p}_y - \hat{q}_y\hat{p}_x = -i/k_0\, \hat{R}_z^{\mathcal{H}_k}. \qquad (11.7c)$$

Compare with (11.2), noting that the 2–1 matrix element of $\hat{R}_x^{\mathcal{H}_k}$ is $\frac{1}{2}(q_y\Delta_k + \Delta_k q_y) = q_y\Delta_k + \partial_{q_y}$ and similarly for $\hat{R}_y^{\mathcal{H}_k}$. It may be somewhat surprising however that the 'wavized' factors $\hat{q}_y$ and $\hat{q}_y$ do *not* commute with $\hat{p}_z$.

Three reasons for **not** accepting this wavization recipe are: *(a)*, multiplicative operators $\hat{q}_x$ and $\hat{q}_y$ are by themselves not self-adjoint under the inner product (10.13) $\mathcal{H}_k$; *(b)* even classically, $\exp v_x\{q_x, \circ\}$ and $\exp v_y\{q_y, \circ\}$ map the momentum variables $\mathbf{p}$ outside their proper optical range $|\mathbf{p}| \leq n$ —recall this is **not** the Heisenberg algebra and such operators are **not** within the Euclidean algebra; and finally, *(c)*, in Sections 13 and 14 we shall present the Lorentz group generators, properly constructed both in $\wp$ and $\mathcal{H}_k$, where the above recipe does not quite work. The difference will be small enough, however, to merit notice.

Our position here is that we should wavize only variables in $\wp$ that are within the Euclidean algebra into self-adjoint operators in $\mathcal{H}_k$. This means that the maps between (7.5)[23] and (11.2) that should complement the momentum wavization (11.6), are:[24]

$$R_x = q_y h \overset{W}{\mapsto} -\frac{i}{k_0}\hat{R}_x^{\mathcal{H}_k} = -\frac{i}{k_0}\begin{pmatrix} 0 & q_y \\ -q_y\Delta_k - \partial_{q_y} & 0 \end{pmatrix}, \quad (11.8a)$$

$$R_y = -q_x h \overset{W}{\mapsto} -\frac{i}{k_0}\hat{R}_y^{\mathcal{H}_k} = -\frac{i}{k_0}\begin{pmatrix} 0 & -q_x \\ q_x\Delta_k + \partial_{q_x} & 0 \end{pmatrix}, \quad (11.8b)$$

$$R_z = \mathbf{q}\times\mathbf{p} \overset{W}{\mapsto} -\frac{i}{k_0}\hat{R}_z^{\mathcal{H}_k}$$
$$= -\frac{i}{k_0}\begin{pmatrix} q_x\partial_{q_y} - q_y\partial_{q_x} & 0 \\ 0 & q_x\partial_{q_y} - q_y\partial_{q_x} \end{pmatrix}. \quad (11.8c)$$

In quantum mechanics, *position* is a very good observable: its eigenstates

---

[22]In quantum mechanics, if two observables quantize as $a \overset{Q}{\mapsto} \hat{A}$ and $b \overset{Q}{\mapsto} \hat{B}$, the symmetrization scheme in entails that the product quantize through their *anticommutator*: $ab \overset{Q}{\mapsto} \frac{1}{2}\{\hat{A}, \hat{B}\}_+ = \frac{1}{2}(\hat{A}\hat{B} + \hat{B}\hat{A})$. When ab is of the form $qf(p)$, $pf(p)$, or quadratic in q and p, this scheme is equivalent to any other quantization scheme [12].

[23]Note that the 'classical' functions to be wavized are *minus* the functions that appear in the Poisson operator.

[24]Although $h = p_z$, we write h in place of p_z here to emphasize that the components of $\mathbf{q}$ will not be wavized alone, but only *in company* with the h's.

are Dirac δ's that, while not quite in $\mathcal{L}^2(\Re)$, are nevertheless limit points of weak sequences of functions that are. One feature of Euclidean-based wavization is the absence of a good position operator in Helmholtz optics. Dirac δ's are nowhere near to functions in the space. The coordinates appear only as the arguments of functions in $\mathcal{H}_k$ and are not extractable from there as eigenvalues of a polynomial operator. Correspondingly, in $\mathcal{H}_k$ The closest we can come to "screen coordinates" are the *rotation* functions $q_x h$ and $q_y h$ above.[25] These functions do not have zero Poisson brackets nor do their operators commute, for they are generators of a rotation group $\mathcal{R}_3$. Educated intuition confirms that they ***should not*** be simultaneously observable, since $\omega(\mathbf{q})$ and $\varpi(\mathbf{q})$ are the 'sharpest' screen functions available. Indeed, we expect to find a form of the sampling theorem (valid for the *sphere*, rather than the circle or torus, as is usual in power spectrum and signal theory [10]). Such is a good program to be followed elsewhere.

6.12 The ray direction sphere under Lorentz boost transformations

It is natural to follow the strategies of classical and quantum mechanics in developing Euclidean optics. The formulation of mechanics may be derived from the nilpotent Heisenberg-Weyl [11], its enveloping algebra, group, and ring [12], from the Galilei group [13], or from the general symplectic group [14]. These groups are inappropriate for optics because here the *momentum* observable has a *bounded* range, whereas in mechanics it is infinite. The issue of *position* coordinates has arisen above and we have seen that they are outside the pale of good Euclidean operators. *Linear* canonical transformations of phase space, a well-studied terrain common to classical and quantum mechanics, are thus meaningless in global optics. Yet because they constitute the essence of the paraxial approximation and the tool of aberration expansions we shall continue to work with them in the near-metaxial regime –elsewhere. Here we now review a group of transformations that is beyond the Euclidean group but whose action *is* well-defined and *global* both in geometric and Helmholtz optics, and follows the wavization process proposed above.

In geometric optics the basic object, a light ray, is a coset $\{\mathbf{p}, \sigma; \mathbf{q}\}$ in the Euclidean group by the symmetry group of the ray. In Helmholtz optics, the basic object is a plane and its corresponding space of cosets has been divided into irreducible subspaces $\{\mathbf{p}; \sigma\}|_k$. They have in common the ray direction sphere $\vec{p} \in \mathcal{S}_2$. The sphere is a well-known subject of relativistic Lorentz SO(3,1) transformations because it is a space of cosets of that

---

[25] We are not allowed to divide by h, because the spectrum of $\hat{h} = -i/k_0\, \hat{T}_z^{\mathcal{H}_k}$ includes zero.

group by the noncompact factor. Physically, this comes about as follows.

Let $\ell = (\vec{\ell}, \ell_0) = (\ell_x, \ell_y, \ell_z, \ell_0)$ be a lightlike four-vector, $|\vec{\ell}| = |\ell_0|$, undergoing a Lorentz boost by $v = c \tanh \alpha$ in the z–direction,

$$\begin{aligned} \ell_{x,y} &\mapsto \ell'_{x,y}, && (12.1a)\\ \ell_z &\mapsto \ell'_z = \ell_z \cosh\alpha + \ell_0 \sinh\alpha, && (12.1b)\\ \ell_0 &\mapsto \ell'_0 = \ell_z \sinh\alpha + \ell_0 \cosh\alpha. && (12.1c) \end{aligned}$$

The *direction* of such a four-vector on a sphere $\mathcal{S}_2$ of radius n is given by the components of $\vec{\ell}$ normalized by division through ℓ_0/n, namely,

$$\vec{p} = (p_x, p_y, p_z) = \left(\frac{n\ell_x}{\ell_0}, \frac{n\ell_y}{\ell_0}, \frac{n\ell_z}{\ell_0}\right). \tag{12.2}$$

In these *homogeneous* coordinates, the boost (12.1) becomes the transformation

$$\mathbf{p} \mapsto \mathbf{p}' = \frac{\mathbf{p}}{\cosh\alpha + p_z/n \,\sinh\alpha}, \tag{12.3a}$$

$$p_z \mapsto p'_z = \frac{p_z \cosh\alpha + n \sinh\alpha}{\cosh\alpha + p_z/n\, \sinh\alpha}, \tag{12.3b}$$

where $\mathbf{p} = (p_x, p_y)$ as usual, and $p = |\mathbf{p}|$. In terms of angles $\{\theta, \phi\}$ over the sphere, [*cf.* (3.5*b*)], we find ϕ to be invariant while the colatitude θ follows the nonlinear transformation given by

$$\frac{p}{p_z + n} = \tan\tfrac{1}{2}\theta \mapsto \tan\tfrac{1}{2}\theta' = e^{-\alpha} \tan\tfrac{1}{2}\theta. \tag{12.4}$$

An observer in a spacecraft moving with respect to the stars will therefore see the directions of their rays concentrate towards his direction of motion by the amount (12.4). This effect was noticed in 1725 by Bradley, who termed it *stellar aberration*, recognized it to originate from the earth's orbital motion during the year, and provided the first estimate of the speed of light. It is a global *deformation*[26] transformation of $\mathcal{S}_2$ that has a group-theoretic origin.

To find the z–boost *generator* responsible for the nonlinear transformation (12.4), we may linearize it to a translation through the change of variables

$$\zeta = -\ln\tan\tfrac{1}{2}\theta \mapsto \zeta' = \zeta + \alpha = \exp\left(\alpha\frac{d}{d\zeta}\right)\zeta. \tag{12.5}$$

From $e^{-\zeta} = \tan\tfrac{1}{2}\theta = \dfrac{p}{n + \sqrt{n^2 - p^2}}$, we find

$$p_\pm = n\frac{2e^\zeta}{1 \pm e^{2\zeta}}, \qquad p_z = n\frac{e^{2\zeta} - 1}{1 + e^{2\zeta}}. \tag{12.6}$$

[26] *i.e.*, the measure over the sphere (10.1) is *not* preserved [15].

On the $\mathbf{p}$–disk δ_n there are *two* values of p for each value of ζ, reflecting the map $\mathcal{S}_2 \xmapsto{2:1} \delta_n$. The boost generator $\hat{B}_z$ effecting $\exp(\alpha \hat{B}_z) f(\zeta) = f(\zeta+\alpha)$, is thus

$$\hat{B}_z = \frac{d}{d\zeta} = -\frac{p_z\, p}{n} \frac{d}{dp} = \frac{\mp\sqrt{n^2-p^2}}{n} \mathbf{p} \cdot \frac{\partial}{\partial \mathbf{p}}. \tag{12.7}$$

6.13 Relativistic coma in geometric optics

When images are enlarged as in a slide projector, or reduced as in a camera, the angles of the rays that arrive at the screen to form the image are inversely reduced or enlarged. This well-known property of passive optical devices is succintly described by the statement that optical transformations must preserve the phase space volume element $d\mathbf{p}\, d\mathbf{q}$. That is, they are bound to produce only *canonical* transformations of the coset manifold seen in Sections 6 and 7. The relativistic Lorentz transformation (12.3) of ray directions, expanded in series of $|\mathbf{p}|$ to fifth order, is

$$\mathbf{p} \mapsto \mathbf{p}' = \frac{\mathbf{p}}{\cosh\alpha + h/n \, \sinh\alpha} \tag{13.1a}$$

$$= e^{-\alpha}\mathbf{p} + \tfrac{1}{2} n^{-2} \sinh\alpha \, e^{-2\alpha} p^2 \mathbf{p} \tag{13.1b}$$

$$+ \tfrac{1}{4} n^{-4} \sinh\alpha \, e^{-2\alpha} (1 - \tfrac{1}{2} e^{-2\alpha}) (p^2)^2 \mathbf{p} + \cdots. \tag{13.1c}$$

As before we abbreviate $h = \pm\sqrt{n^2 - p^2}$, the sign indicating the hemisphere and $p^2 = p_x^2 + p_y^2$. To first order in p, we have a magnification by a factor of $e^{-\alpha}$ that is less than unity for $\alpha > 0$; after this we have the series of terms that tell us that the magnification is not linear, but a *distortion* of the ray direction sphere. Therefore, to first order in the ray *position* $\mathbf{q}$ (intersection with the standard screen), we expect a magnification with the inverse factor e^{α}. This will be followed by *aberration* of the nature of *coma*, as will be borne out below.

In order to *extend* the boost action (12.6) from the direction sphere $\vec{p}$ to the whole space $\wp$ of geometric optics $\{\mathbf{p}, \sigma; \mathbf{q}\}$ *canonically*, we note that the boost generator in (12.7) may be written as

$$\hat{B}_z = -\frac{h}{n} \mathbf{p} \cdot \frac{\partial}{\partial \mathbf{p}} = \frac{\partial(-h\mathbf{p}\cdot\mathbf{q}/n)}{\partial \mathbf{q}} \cdot \frac{\partial}{\partial \mathbf{p}}. \tag{13.2a}$$

This is the $\partial/\partial\mathbf{p}$ part of a Poisson operator [*cf.* Eq. (7.6)], and suggests we *extend* it to $\wp$ as

$$\hat{B}_z^{\wp} = \{-h\mathbf{p}\cdot\mathbf{q}/n, \circ\} = -\frac{h}{n}\mathbf{p} \cdot \frac{\partial}{\partial \mathbf{p}} + \frac{h}{n} \mathbf{q} \cdot \frac{\partial}{\partial \mathbf{q}} - \frac{\mathbf{p}\cdot\mathbf{q}}{nh} \mathbf{p} \cdot \frac{\partial}{\partial \mathbf{q}}. \tag{13.2b}$$

Now we can exponentiate this operator and show[27] that the boost action on ray position at the screen, conjugate to (13.1), is

$$\begin{aligned} \mathbf{q} &\mapsto \mathbf{q}' \\ &= \exp(\alpha \hat{B}_z^{\wp})\,\mathbf{q} && (13.3a) \\ &= (\cosh\alpha + h/n\,\sinh\alpha)\left(\mathbf{q} - \frac{\sinh\alpha}{n\sinh\alpha + h\cosh\alpha}\,\frac{\mathbf{p}\cdot\mathbf{q}}{n}\,\mathbf{p}\right) && (13.3b) \\ &= e^{\alpha}\mathbf{q} - n^{-1}\sinh\alpha\,\mathbf{p}\cdot\mathbf{q}\,\mathbf{p} - \tfrac{1}{2}n^{-2}\sinh\alpha\,p^2\mathbf{q} \\ &\quad - \tfrac{1}{2}n^{-3}\sinh\alpha\, e^{-2\alpha}p^2\mathbf{p}\cdot\mathbf{q}\,\mathbf{p} - \tfrac{1}{8}n^{-4}\sinh\alpha\,(p^2)^2\mathbf{q} \\ &\quad - \cdots. && (13.3c) \end{aligned}$$

By construction it is guaranteed that the measure $d\mathbf{p}\,d\mathbf{q}$ will be preserved.[28] In the last expression we have developed the closed formula (13.3*a*) in series of powers of $|\mathbf{p}|$ and $|\mathbf{q}|$ to fifth order. The leading linear term shows indeed a magnification factor of e^{α} as required. The rest of the series contains terms in $(p^2)^m\mathbf{q}$ and $(p^2)^{m-1}\mathbf{p}\cdot\mathbf{q}\,\mathbf{p}$, $m = 1, 2, \ldots$. The presence of only such terms determines the mapping to be circular *comatic*. This is the name of a class of aberrations that are *comet*-shaped, and that have the very important property of being 2:1 mappings of object rays to screen points: $(\mathbf{p},\mathbf{q})$ and $(-\mathbf{p},\mathbf{q})$ are mapped on the *same* image *point* $\mathbf{q}'(\mathbf{p},\mathbf{q})$.[29]

Lorentz boosts in directions other than the screen normal may be obtained transforming the generator $\hat{B}_z^{\wp}$ in (13.2*b*) by means of the generators of rotations given in (7.5) through Poisson operators. In this way we find

$$\begin{aligned} \hat{B}_x^{\wp} &= \{nq_x - p_x\,\mathbf{p}\cdot\mathbf{q}/n, \circ\} = \{q_x h^2 + p_y\,\mathbf{q}\times\mathbf{p}, \circ\}, && (13.4a) \\ \hat{B}_y^{\wp} &= \{nq_y - p_x\mathbf{p}\cdot\mathbf{q}/n, \circ\} = \{q_y h^2 - p_x\,\mathbf{q}\times\mathbf{p}, \circ\}, && (13.4b) \end{aligned}$$

where $\times$ is the vector cross product. The three-vector of boosts is thus generated by the Poisson operator of the vector function $\vec{B} = \vec{p}\times\vec{R}$, where $\vec{R} = \vec{q}\times\vec{p}$ with $\vec{q} = (\mathbf{q}, 0)$ is the three-vector function generating rotations through Poisson operators given in (7.1)–(7.5).[30] Since these are three-vectors, it is natural to expect that

$$[\hat{B}_i^{\wp}, \hat{R}_j^{\wp}] = -\varepsilon_{ijk}\hat{B}_k^{\wp} \qquad (13.5a)$$

holds, as it does. Moreover, it is *also* true that

---

[27] The way to derive this formula will be indicated below. *Prima facie*, this is a nontrivial task.

[28] The Poisson bracket of the transformed variables (13.1*a*)–(13.3*a*) is also conserved: $\{q_i', p_j'\} = \delta_{i,j} = \{q_i, p_j\}$, etc.

[29] We must emphasize the word *point* because the two image *rays* are distinguished by their *direction* at the screen. Witness in (13.1) that $\mathbf{p}'(\mathbf{p},\mathbf{q}) = -\mathbf{p}'(-\mathbf{p},\mathbf{q})$, as required by the essential 1:1 bijection of all *canonical* mappings of phase space.

[30] Note that $\{\vec{q}, \vec{p}, \vec{R}\}$ form a right triad of vectors.

$$[\hat{B}_i^{\wp}, \hat{B}_j^{\wp}] \quad = \quad +\varepsilon_{ijk}\hat{R}_k^{\wp}. \tag{13.5b}$$

Hence the $\hat{R}^{\wp}$'s and $\hat{B}^{\wp}$'s close into the algebra of the Lorentz SO(3,1) group of special relativity.[31] It is left as an exercise to the reader to decide whether this fact is natural or remarkable. We also note the x–y vector identity $\mathbf{q} = \mathbf{b}/n - b_z\mathbf{p}/nh$, that allowed us to derive the rather formidable Lie exponential of $\hat{B}_z^{\wp}$ in (13.3) through knowing the transformation properties of the pieces in the right-hand side. Not so obvious is the boost exponential in the x–direction, that may be shown to be[32]

$$\begin{aligned} \exp(\alpha\hat{B}_x^{\wp})\, q_x &= q_x' = (\cosh\alpha + p_x/n\, \sinh\alpha) \\ &\qquad \times (q_x \cosh\alpha + \mathbf{p}\cdot\mathbf{q}/n\, \sinh\alpha), \qquad (13.6a) \\ \exp(\alpha\hat{B}_x^{\wp})\, q_y &= q_y' = (\cosh\alpha + p_x/n\, \sinh\alpha)\, q_y. \qquad (13.6b) \end{aligned}$$

The transformation undergone by p_x and p_y may be found from (12.3) with the rotated replacement $(p_x, p_y, p_z) \mapsto (p_z, p_x, p_y)$.

Take an image-forming device that is in focus when at rest, and then boost it to α. Figures 5, 6, and 7 show[33] respectively what our mathematics predicts should be the image formed by an array of luminous points on a screen moving towards $(+z)$, sideways $(+x)$, and away from $(-z)$ the optical axis, at the rather considerable speed of $\alpha = 0.3$ ($v = 0.29131\,c$). The images are supposed to be formed out of a 45°–cone of directions around the forward pole (optical axis). The paralells and meridians of this spherical cap constitute the *spot diagram* of the original point images (marked by crosses) magnified and aberrated by the Lorentz motion represented by equations (13.1)–(13.3) and (13.6).

Some detailed geometric properties of the figures have been explored in reference [16]. Here we only want to remark that the three figures are faces of the same aberration, and that they are global: they appear as circular comatic for small angles in $\pm z$–motion and astigmatic/curvature of field for

---

[31] Posed as a group deformation procedure, we may add to $\vec{B}$ a multiple μ of $\vec{p}$ and still have the commutation relations (13.5) close. The Lorentz invariants are $\vec{B}^2 - \vec{R}^2 = n^2\mu^2$ and $\vec{R}\cdot\vec{B} = 0$.

[32] These expressions were found by Wolfgang Lassner by a back-and-forth process involving hand and symbolic REDUCE computation on the trusty old IIMAS/Cuernavaca PC, checking that two successive x–boosts will compose properly.

[33] I would like to thank Guillermo Correa (IIMAS–UNAM/DF) for the graphics program SPOT_D, that is capable of reading muSIMP output files through PASCAL and plotting the corresponding spot diagrams; it works not only with aberration expansions, but with exact *global* formulas that apply for Euclidean optics. It is reported in: G.J. Correa–Gómez and K.B. Wolf, SPOT_D, *Programa para Graficación de Diagramas de Manchas en Optica.* Comunicaciones Técnicas IIMAS, Serie Desarrollo, No. 97 (1989), 51 págs. The program is open and may be requested from the author.

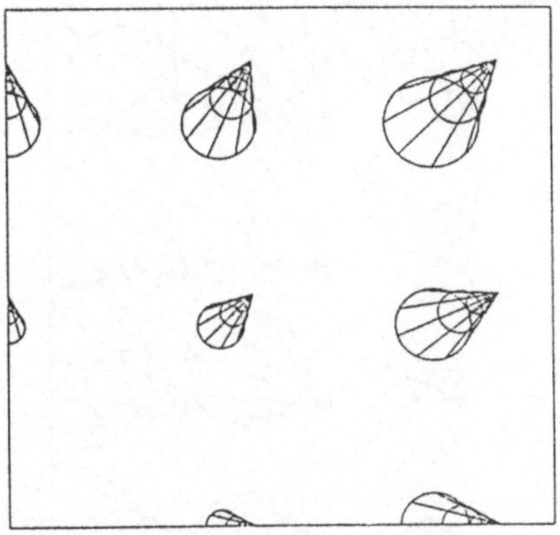

FIGURE 5. Relativistic coma in geometric optics. A screen receives the focused image of an array of object points through collecting rays from 45° cones. When the screen approaches the source at a velocity of $v = 0.29131\,c$ ($\alpha = 0.3$) the image amplifies and exhibits global coma.

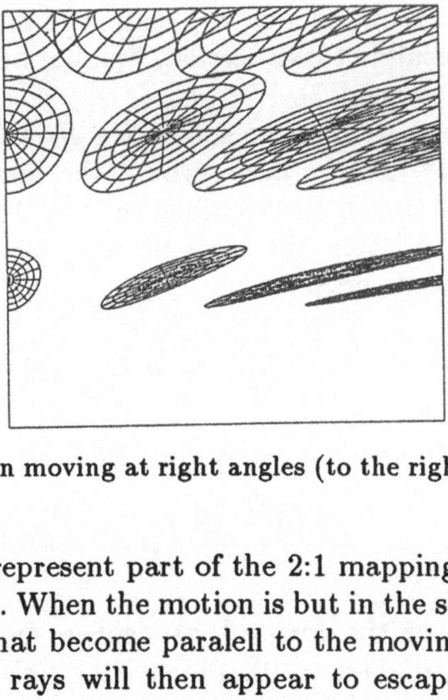

FIGURE 6. The screen moving at right angles (to the right) at the same velocity.

cross motion; they represent part of the 2:1 mapping of the *full* Descartes sphere on the screen. When the motion is but in the screen plane, there will be a circle of rays that become paralell to the moving screen; the position coordinate of these rays will then appear to escape to infinity, without implying any actual singularity in the ray manifold.

6.14 Relativistic coma in Helmholtz wave optics

Helmholtz optics was presented in Sections 9 to 11 as the Euclidean geometry of planes belonging to a definite irreducible representation k of that group. Here we want to explore the Helmholtz wave optics representation of the relativistic Lorentz transformation seen in Section 12. We shall do

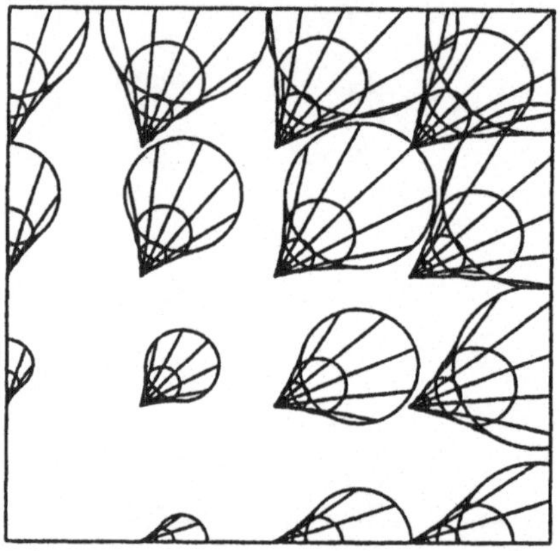

FIGURE 7. The screen moving away from the object. Figures 5 and 7, when superposed, show the $\theta < 45°$ and $\theta > 135°$ parts of global coma.

this first 'longhand', by conventional means of the group deformation arguments of Ref. [15], and then through application of the wavization process (11.6)–(11.8) on the results of last Section. Finally, we comment upon the results obtained in reference [17].

The Lorentz transformation of the sphere $\vec{p} \in \mathcal{S}_2$ was shown to have the formal generator (12.7). Taking into account the measure of the $\mathcal{L}^2(\mathcal{S}_2)$ inner product (10.9*a*)–(10.9*b*), and recalling the matching condition among the two functions of the latter, the *skew-adjoint* generator of z–boosts on $\mathcal{L}^2(\mathcal{S}_2)$ is its symmetrized version,

$$\hat{B}_z^{\mathcal{S}_2} = -\frac{p_z}{2n}\left(\mathbf{p}\cdot\frac{\partial}{\partial\mathbf{p}} + \frac{\partial}{\partial\mathbf{p}}\cdot\mathbf{p}\right), \tag{14.1a}$$

and the two-chart operator for $\mathcal{H}_k$ over the disk δ_n is

$$\hat{B}_z^{k\pm} = \frac{\mp\sqrt{n^2-p^2}}{n}\left(\mathbf{p}\cdot\frac{\partial}{\partial\mathbf{p}} + 1\right). \tag{14.1b}$$

The upper sign applies on the 'forward' hemisphere ($p_z > 0$) and the lower sign in the 'backward' one ($p_z < 0$). Finally, through replacement in the integrand of the transform pair (10.5), integration by parts (with the appropriate cancellation of boundary terms for Φ_+ with Φ_-) and extraction from the integral, we obtain the z–boost generator skew-adjoint on the Helmholtz Hilbert space $\mathcal{H}_k$ on the screen:

$$\hat{B}_z^{\mathcal{H}_k} = \frac{i}{k}\begin{pmatrix} 0 & -\hat{D} \\ (\hat{D}+1)\Delta_k - k^2 & 0 \end{pmatrix}, \tag{14.2}$$

where $\hat{D} = \frac{1}{2}(\mathbf{q}\cdot\partial_\mathbf{q} + \partial_\mathbf{q}\cdot\mathbf{q})$ is, as in (11.3) the usual dilatation operator, and Δ_k the Laplacian on the screen plus k^2.

This Lorentz z–boost generator can be written abstractly in terms of the Euclidean generators $\hat{T}_j$ and $\hat{R}_j$, $j = x, y, z$ given in (9.3)–(9.4), (11.1)–(11.2), or any other realization where $\sum_j \hat{T}_j^2 = -k^2$, a constant, in the following forms:

$$\begin{aligned} \hat{B}_z &= \frac{i}{k}(\hat{T}_x \hat{R}_y - \hat{T}_y \hat{R}_x + \hat{T}_z) && (14.3a) \\ &= \frac{i}{k}(\hat{R}_y \hat{T}_x - \hat{R}_x \hat{T}_y - \hat{T}_z) && (14.3b) \\ &= \frac{i}{k}\{\hat{R}_y \hat{T}_x\}_+ - \frac{i}{k}\{\hat{R}_x \hat{T}_y\}_+ && (14.3c) \\ &= -\frac{i}{k}[(\hat{R}_x)^2 + (\hat{R}_y)^2 + (\hat{R}_z)^2, \hat{T}_z], && (14.3d) \end{aligned}$$

where $\{A, B\}_+ = AB + BA$ is the anticommutator, $[A, B] = AB - BA$ the commutator. All operators are here skew-adjoint. The last form (14.3d) allows us to write the vector boost generator as

$$\hat{\vec{B}} = \frac{i}{k}[\hat{R}^2, \hat{\vec{T}}]. \tag{14.4}$$

This is the usual algebra deformation formula for ISO(3) $\Rightarrow$ SO(3,1) [15], and insures that the three components of $\hat{B}_j$ close, together with the rotation generators $\hat{R}_j$, into the Lorentz algebra:[34]

$$\begin{aligned} [\hat{B}_i, \hat{R}_j] &= -\varepsilon_{ijk}\hat{B}_k, && (15.4a) \\ [\hat{B}_i, \hat{B}_j] &= +\varepsilon_{ijk}\hat{R}_k. && (15.4b) \end{aligned}$$

In $\mathcal{L}^2(\mathcal{S}_2)$ the boost generators are first-order skew-adjoint differential operators in the components of $\vec{p}$, whose expression includes the symmetrized version of the $\mathbf{p}$–operator part of the corresponding geometric-optics operators in (13.2) and (13.4). In the Helmholtz Hilbert space $\mathcal{H}_k$ of two-functions, they are 2×2–matrices with $\partial/\partial\mathbf{q}$–operator entries that may be found from (11.1)–(11.2) and (14.3)–(14.4). The results so obtained were reported in reference [17]. They are given by (14.2) for $\hat{B}_z^{\mathcal{H}_k}$ and, in screen two-vector form,

$$\hat{\mathbf{B}}^{\mathcal{H}_k} = -\frac{i}{k}\begin{pmatrix} \hat{D}\partial_{\mathbf{q}} + k^2\mathbf{q} & 0 \\ 0 & (\hat{D}+1)\partial_{\mathbf{q}} + k^2\mathbf{q} \end{pmatrix}. \tag{14.6}$$

The same results for the boost generators may be obtained through applying the *wavization* rules given in (11.6)–(11.8) on the function compo-

[34] We are again allowed to add any real multiple μ of $\hat{\vec{T}}$ to (14.4), leading to all nonexceptional continuous series of SO(3,1) representations according to the deformation algorithm. The values of the Casimir operators are $\hat{R}^2 - \hat{B}^2 = 1$ and $\hat{R} \cdot \hat{B} = 0$:

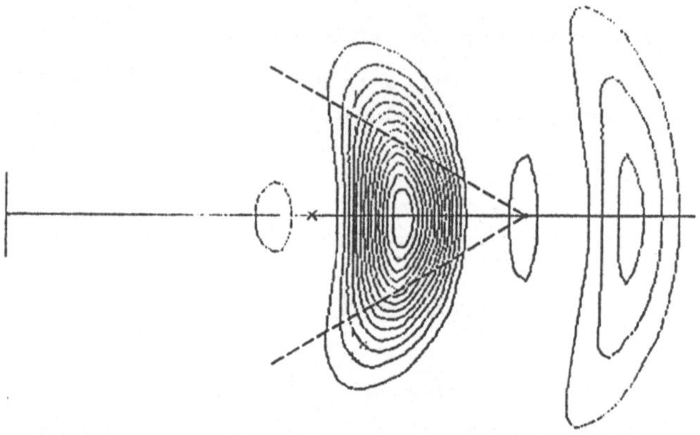

FIGURE 8. Relativistic coma in Helmholtz optics. A Gaussian beam displaced from the optical center by $2/k_0$ units (marked by the $\times$) and of width $4/k_0^2$ units is focused on a screen. When the screen approaches the source at $v = 0.29131\,c$ ($\alpha = 0.3$), as in Figure 5, the image exhibits the indicated 'isophote' lines given by the square of the first component of the Helmholtz two-function.

nents of the three-vector

$$\vec{B} = \vec{p}\times\vec{R} = \begin{pmatrix} p_z\, q_x h + p_y\, \mathbf{q}\times\mathbf{p} \\ -p_x\, \mathbf{q}\times\mathbf{p} + p_z\, q_y h \\ -p_x\, q_x h - p_y\, q_y h \end{pmatrix}, \tag{14.7}$$

where we write the components as a column to save space. They contain the same *functions* that appear in the Poisson operators (13.2*b*)–(13.4) for geometric optics. We emphasize that for the purpose of wavization, we must consider $q_x h$ and $q_y h$ as *single* observable subject to the wavization mapping. If we were to replace hp_z by $p_z^2 = n^2 - p_x^2 - p_y^2$ in the first two components of $\vec{B}$, we would obtain $n^2\mathbf{q} - \mathbf{p}\cdot\mathbf{q}\,\mathbf{p}$ —tempting us to wavize $\mathbf{q}$ alone. The wavization of *this* form of the function would be a *diagonal* matrix with equal elements $(\hat{D} + \frac{1}{2})\partial_{\mathbf{q}} + k^2\mathbf{q}$.[35] The z–component of $\vec{B}$, on the other hand, involves only the combinations $h\mathbf{q}$, and its wavization yields the correct result anyway.

Let us now report on the essentials of the exponentiation of the z–boost generator (14.2), $\exp(i\alpha\hat{B}_z^{\mathcal{H}_k})$, carried out in reference [17]. Let us prominently note that the 1–2 matrix element contains $\hat{D}$, by itself the generator of magnifications $\exp(\alpha\hat{D}) : f(\mathbf{q}) \mapsto e^{\alpha/2} f(e^{\alpha}\mathbf{q})$. The 2–1 matrix element contains the inverse magnification plus the Schrödinger-quantized *coma-*generating function $p^2\, \mathbf{p}\cdot\mathbf{q}$ [18].

In reference [17] we expanded the matrix operator $\hat{B}_z^{\mathcal{H}_k}$ in series to fifth order in α, involving differential operators up to degree nine. This was applied with a symbolic computation `muSIMP` program to a *forward* Gaussian beam with waist at the screen, *off* the optical axis. Strictly, of course, a

[35] This is, in fact the *average* of the two diagonal matrix elements in (14.6).

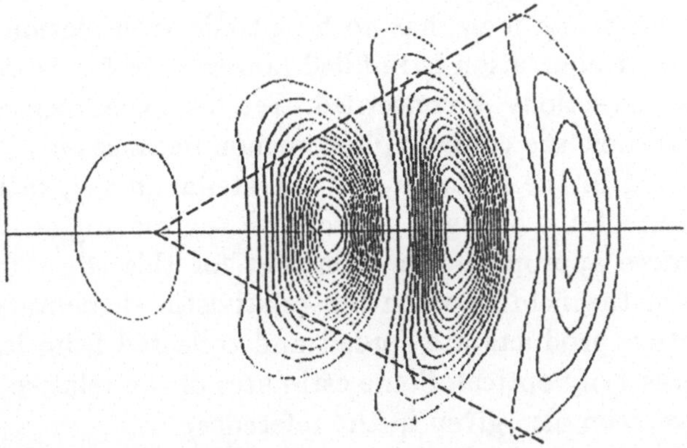

FIGURE 9. The same beam focused on a screen receding from the source at that velocity. The dashed lines indicate the geometric Seidel coma caustics at the apex. The position of the apex is shifted due to magnification and reduction for each case.

Gaussian function is not in the Helmholtz Hilbert space $\mathcal{H}_k$ because its Fourier transform has small but nonzero support outside the disk δ_n. Calculational expedience, however, makes such a beam an irresistible candidate for investigation: successive ∂_{q_j}–derivatives simply pile a q-polinomial factor in front of it. The resulting Lorentz-transformed squared function was then evaluated on a square grid, and the numerical matrix fed into a plotting algorithm that drew the spline level curves shown in Figure 8, for $\alpha = +0.3$ and Figure 9 -0.3, respectively.[36] These figures should be compared with the ones for the geometric optics phenomenon seen in the last Section. The geometric coma caustic angle (of 60°) and origin (shifted by e^{α}) is superposed on the figures here. With increasing truncation order of α, the single Gaussian peak unfolds into local maxima, separated by an increasing number of crescent-shaped dark fringes, whose overall features seem to stabilize by degree five in α. Comparing the figures with the pattern of diffraction in coma aberration [19], it would seem that the $\alpha = -0.3$ figure comes closer to that of pure coma than the $\alpha = +0.3$ figure. Actually, the series contains also, prominently, the magnification generator that contributes itself with circular fringes centered on the optical axis. These counteract the curved fringes of pure coma in the first figure, and reinforce those in the second.

The relativistic coma phenomenon predicted for geometric and Helmholtz optics is in a sense perplexing, since none of the two models entertains

---

[36]I would like to thank José Fernando Barral, of the Instituto de Astronomía, UNAM, for the indispensable help with the figures in this Section. Some ideas on applications to digital image processing were also aired and may be pursued with José Luis Morales, at the Instituto Nacional de Astrofísica, Optica y Electrónica, Tonantzintla.

a *time* variable and, *prima facie*, has nothing to do with motion. Bradley's observation of stellar aberration is fulfilled however, as far as mappings of the sphere of ray directions is concerned. Yet his experimental setup (a telescope) will *not* show the comatic phenomenon because only a *single* ray direction is involved in the formation of the stellar image, rather than a spread pencil of directions brought to focus on the moving screen. It may well be that a more appropriate environment for this effect to manifest its properties is in the field of radiating relativistic elementary particles whose disintegration products are spread and collected from large angles by means not necessarily optical. Some estimates of the relative *size* of the coma caustic have been also given in the reference.

6.15 Reflection, refraction, and concluding remarks

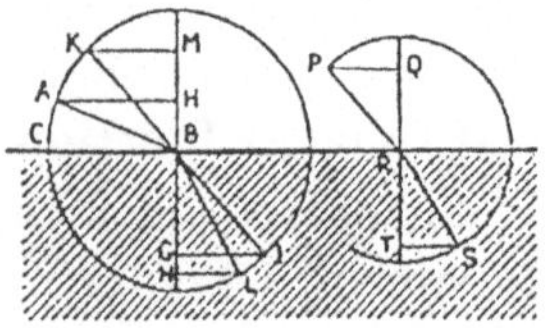

As, for example, if there passes a ray in Air from A *to* B *that finds at* B *the surface of Glass* CBR, *it detours to* I *in this Glass; and another from* K *to* B *detours to* L; *and another from* P *to* R *that detours to* S; *there must be the same proportion between the lines* KM *and* LN, *or* PQ *and* ST, *than between* AH *and* IG, *but not the same between the angles* KBM *and* LBN, *or* PRQ *and* SRT, *than between* ABH *and* IBG.

René Descartes, *Discourse on the Method*
Second Discourse: *On Dioptrics*

Throughout this paper we have considered the radius of the Descartes sphere of ray directions to be a fixed number n; in passing to Helmholtz optics we defined the wavenumber in the medium to be $k = nk_0$, with k_0 the wavenumber for $n = 1$. The need for this unit is not in Euclidean optics of a single homogeneous medium, but to allow for the phenomenon of refraction, where at least *two* homogeneous media are involved. This we do here starting with reflections, and ending with some concluding remarks on what has been done to establish a theory of global optics with wavization, and what remains to be done.

In geometric optics, we recall from Section 5, a single light ray is a coset in the Euclidean group $\{\mathbf{p}, \sigma; \mathbf{q}\}$, a point in the space $\wp = \mathcal{H}^{\text{geom}} \backslash \mathcal{E}_3$. In Helmholtz optics, Sections 9 and 10, a single plane wave is the irreducible component $k = nk_0$ of a coset in $\mathcal{H}^{\text{wf}} \backslash \mathcal{E}_3$, characterized by $\vec{p}$ or $\{\mathbf{p}, \sigma\}$ [see Eqs. (10.7)], and realized by the two-function $\mathbf{W}_{\mathbf{p},\sigma}(\mathbf{q})$ (amplitude and normal derivative) at the $z = 0$ screen $\mathbf{q} \in \Re^2$ given in (10.8). We may define the following two physical operations on the two-disk projected Descartes sphere of ray directions:

$$\text{Reflection} \qquad \nearrow \mapsto \nwarrow, \qquad \mathsf{R} : \{\mathbf{p}, \sigma\} \mapsto \{\mathbf{p}, -\sigma\}, \qquad (15.1a)$$

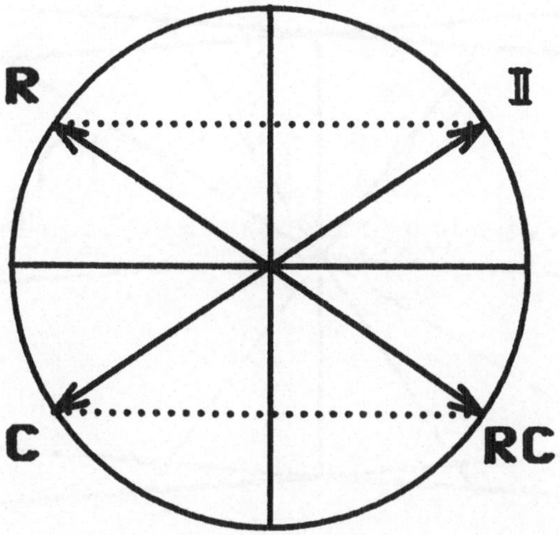

FIGURE 10. Four rays on the Descartes sphere: the original ray **I**, the reflected ray **R**, the conjugate ray **C**, and the reflected-conjugate ray **RC**.

$$\text{Conjugation} \quad \mapsto \quad , \qquad \mathbf{C} : \{\mathbf{p}, \sigma\} \mapsto \{-\mathbf{p}, -\sigma\}, \tag{15.1b}$$

and their product

$$\mapsto \quad , \qquad \mathbf{RC} : \{\mathbf{p}, \sigma\} \mapsto \{-\mathbf{p}, \sigma\}. \tag{15.1c}$$

We have called the two former operations *physical* because the first corresponds to ordinary reflection by a mirror in the $z = 0$ plane, and the second to reflection in a phase-conjugation mirror, that reverses the directions of rays and wavefronts. They both belong to the component of the *orthogonal* group O(3) disconnected from the identity **I**, and may be realized by 3×3 matrices diag$(1, 1, -1)$ and diag$(-1, -1, -1)$, respectively, of determinant -1. Their product $\mathbf{RC} = \mathbf{CR}$ is a proper $\mathcal{R}_3$ rotation by π around the z-axis; also, $\mathbf{R}^2 = \mathbf{C}^2 = \mathbf{I}$. See Figure 10. The small arrow diagrams in (15.1) express our intuition in geometric optics regarding reflections and inversions at the screen. In Helmholtz optics reflection **R** reverses the sign of the second, normal derivative component in (10.8)[37] while conjugation **C** is *complex* conjugation. Note that these operations act exclusively on ray direction, *i.e.*, on points of the Descartes sphere; they do *not* affect **q**, neither in the geometric nor Helmholtz cases.

In geometric optics the operator $\mathbf{I} + \mathbf{C}$ produces a world of *nondirected* " " rays[38] while in Helmholtz optics it leads to purely *real* wavefunctions.

[37]The operator of reflection on the Helmholtz Hilbert space $\mathcal{H}_k$ is realized by the 2×2 matrix diag$(1, -1)$.

[38]We may see them as having Wigner-type distributions that have the same value for $\vec{p}$ and $-\vec{p}$ at every **q**. Alternatively, this property could be introduced

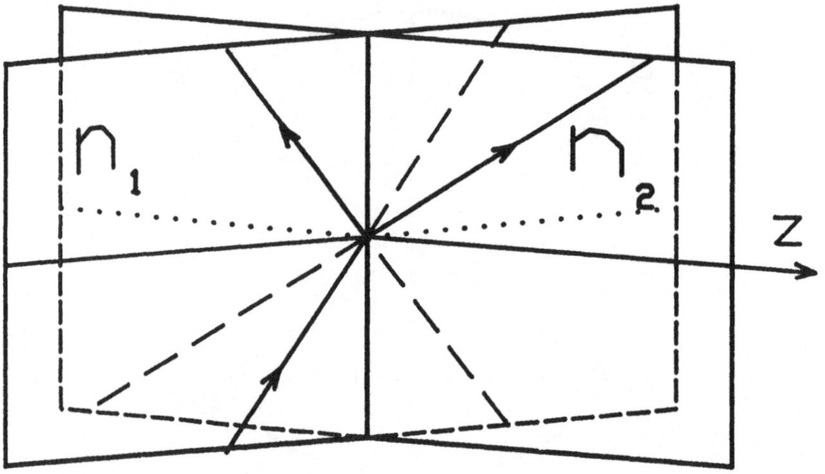

FIGURE 11. The joining of two homogeneous media with two different refractive indices at the reference (1–dim) screen.

The operator **I+R** correspondingly creates a "✳" world where every optical being is accompanied by its reflection by the $z = 0$ screen-turned-mirror; Helmholtz two-functions in particular have zero normal derivative there [see p_z appear in the second component in (10.8); with **I−R** the first component is made to vanish]. We should not conclude that there is "nothing behind the mirror"—whatever that means; only that the distribution of rays or wavefunctions are forced to satisfy certain *boundary conditions* at the $z = 0$ plane. Indeed, through linear combinations of **I**, **C**, **R**, and **CR** acting on distributions on $\wp$ or wavefunctions in $\mathcal{H}_k$, we may describe four-ray $\Re^3$ situations "✳" where the $z = 0$ plane is a special submanifold where boundary conditions can be imposed.

After these reflections, consider *two* homogeneous media, characterized by *different* refractive indices n_1 and n_2, joined at their $z = 0$ plane. We use the word *joined* rather than *separated*, because we have the paradigm of two homogeneous *interpenetrating* media, and postulate that quantities, ray distributions or wavefunctions, match at the reference plane. See the sketch of the two-dimensional analogue in Figure 11.

In geometric optics, let us indicate the coordinates of single rays in the two $\wp$'s by superindices $^{(1)}$ and $^{(2)}$. The *conservation* statements will be

$$\mathbf{q}^{(1)} = \mathbf{q}^{(2)} \in \Re^2, \qquad \delta_{n_1} \ni \mathbf{p}^{(1)} = \mathbf{p}^{(2)} \in \delta_{n_2}, \quad \sigma^{(1)} = \sigma^{(2)}. \quad (15.2a,b,c)$$

---

as part of the $\mathcal{H}^{\mathrm{geom}}$ symmetry group.

or,

$$\wp^{(1)} \ni \{\mathbf{p},\sigma;\mathbf{q}\}^{(1)} = \{\mathbf{p},\sigma;\mathbf{q}\}^{(2)} \in \wp^{(2)} \quad \text{on the common domain.} \tag{15.2d}$$

The conservation of ray position $\mathbf{q}$ at the refracting surface, the equality (15.2a), is a generally implied but not at all irrelevant statement; it is on the same footing as Snell's law of refraction, equality (15.2b). In the common region $\delta_{n_1} \bigcap \delta_{n_2}$, the latter entails $n^{(1)} \sin\theta^{(1)} = n^{(2)} \sin\theta^{(2)}$ and $\phi^{(1)} = \phi^{(2)}$ in the usual spherical coordinates. Equality (15.2c), distinguishes between the two possibilities $\theta^{(2)}$ and $\pi - \theta^{(2)}$. An immediate consequence of (15.2b) is the relation between the z–components of $\vec{p}^{(1)}$ and $\vec{p}^{(2)}$:

$$n_1^2 - (p_z^{(1)})^2 = n_2^2 - (p_z^{(2)})^2. \tag{15.3}$$

The problem for the *global* joining between $\wp^{(1)}$ and $\wp^{(2)}$ is that in the region between the union and the intersection of the δ_n's, one of the p_z's must be imaginary.

A similar set of conservation statements may be made for Helmholtz two-functions in the form

$$\mathbf{F}^{(1)}(\mathbf{q}) = \mathbf{F}^{(2)}(\mathbf{q}), \text{ for all } \mathbf{q} \in \Re^2. \tag{15.4}$$

If we work with the plane waves in (10.8), in linear combination with their reflections through R $C_j^{\nearrow}\mathbf{W}_{\mathbf{p}^{(j)},\sigma^{(j)}}(\mathbf{q}) + C_j^{\nwarrow}\mathbf{W}_{\mathbf{p}^{(j)},-\sigma^{(j)}}(\mathbf{q})$, we find also Snell's law in the form (15.2b) and with the consequence (15.3). There are also two relations for the four linear combination coefficients, $C_1^{\nearrow} + C_1^{\nwarrow} = C_2^{\nearrow} + C_2^{\nwarrow}$ and $p_z^{(1)}(C_1^{\nearrow} + C_1^{\nwarrow}) = p_z^{(1)}(C_2^{\nearrow} + C_2^{\nwarrow})$. The assumption that a transmitted wave is not accompanied by its transmitted reflection provides the ratio of the coefficients of transmission and reflection. As is also well known, this exercise provides the interpretation of imaginary p_z's as *evanescent* waves, exponentially decreasing beyond the boundary where total internal reflection occurs [19]. Such solutions are outside the Hilbert space $\mathcal{H}_k$. Four-wave situations are of interest in the case of dynamic holograms, but this would take us beyond the intended scope of this monograph.

In Figures 12, 13, and 14 we present the three subcases of global refraction for $n_1 < n_2$, between air to the left and glass to the right, say. For $\sigma_1 = +1 = \sigma_2$ rays, we have the traditional rendering in Fig. 12 of refraction into denser media, in company with some reflection back into air that is indicated by dotted lines. In Fig. 13 we picture a $\sigma_1 = -1 = \sigma_2$ ray, for which the same configuration applies with $z = 0$ serving as interface for *right-to-left* glass-to-air refraction, entailing some reflection back into glass. Beyond $\sin\theta^{(2)} = n_1/n_2$, the conservation laws (15.2) *et seq.* imply total internal reflection, as for the ray in Fig. 14, with the presence of an evanescent wave of imaginary p_z lying on one of the branches of the equilateral hyperboloid extending beyond the inner sphere. The three processes occur in global refraction. To the right, corresponding Feynman diagrams

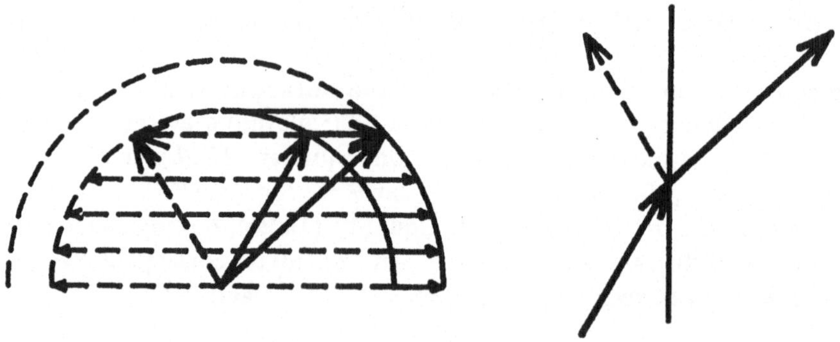

FIGURE 12. The diagram of Descartes for refraction from one lighter medium into one denser.

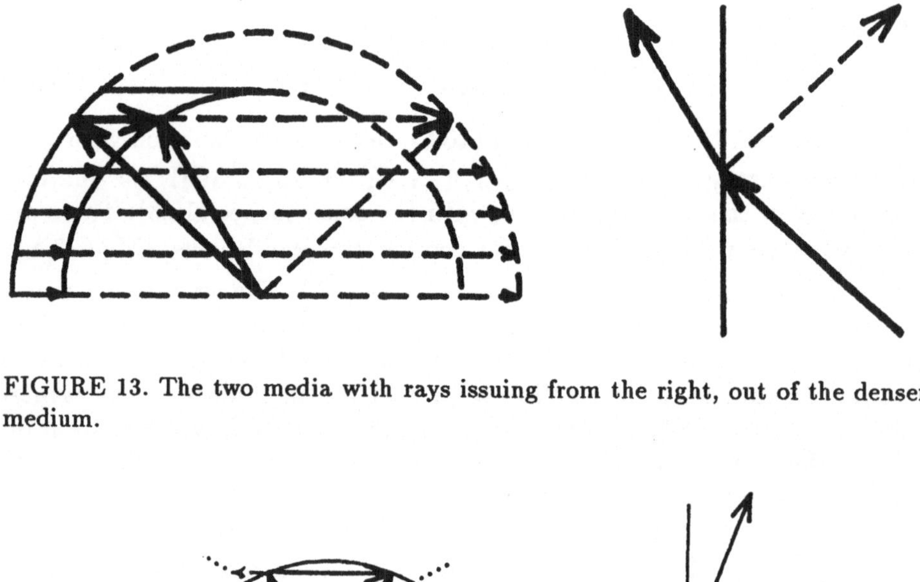

FIGURE 13. The two media with rays issuing from the right, out of the denser medium.

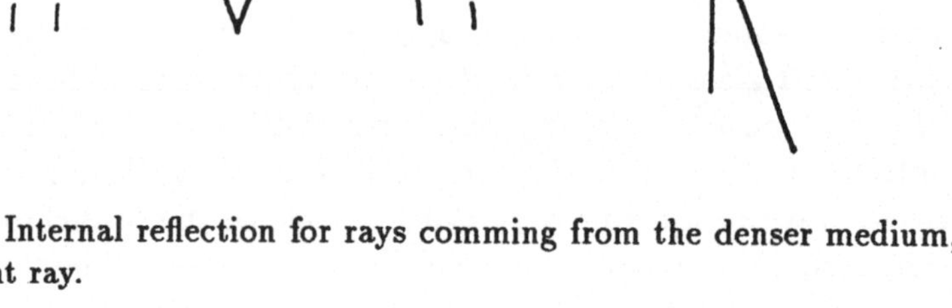

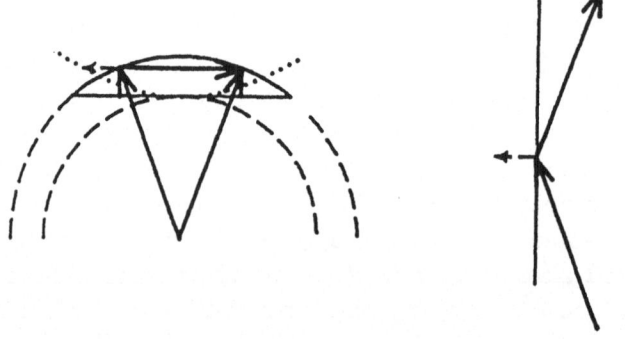

FIGURE 14. Internal reflection for rays comming from the denser medium, and the evanescent ray.

are drawn, borrowing the implications of a field theory on the coset spaces of the Euclidean group.

Refraction through curved surfaces $z = \zeta(\mathbf{q})$, the classical problem of geometrical optics still has to be understood in Euclidean optics. The *root* transformation described in Ref. [18] is amenable to Euclidean treatment, but the transformation depends on the point of contact $\vec{r} = (\bar{\mathbf{q}}, \zeta(\bar{\mathbf{q}}))$ in an implicit form, a function of the ray coordinates $\{\mathbf{p}, \sigma; \mathbf{q}\}$. The problems of rays tangent to the surface and multiple refraction then appear. A telescope, in global optics, receives rays from both ends. This inevitably limits the actual calculational possibilities of Euclidean optics. It is quite clear that for design purposes, we must resort to the well-tried paraxial approximation and the aberration expansion tools developed before.

We must thus point out the following avenue of the Euclidean theory to incorporate lenses, that can be best explained in geometric optical terms. Euclidean operations that are well defined both in Euclidean and in Heisenberg-Weyl paraxial optics [see (7.4)–(7.5)] are x– and y–translations, and z–rotations, generated by $\{\mathbf{p}, \circ\}$ and $\{\mathbf{p} \times \mathbf{q}, \circ\}$. These paraxial $\mathcal{E}_2$ transformations are complemented by x– and y–rotations, generated by $\{-q_y h, \circ\}$ and $\{q_x h, \circ\}$, or by the boosts (13.4), when they are approximated by $\{\mathbf{q}, \circ\}$ through $h = -n + \cdots$. The approximation of z–translation (7.4c) through $h = -n + p^2/2n + \cdots$ yields first-order spherical aberration $\{p^2, \circ\}$ and the z–boost (13.2b) approximates the paraxial magnification generator $\{\mathbf{p} \cdot \mathbf{q}, \circ\}$. However, we do *not* have a Euclidean or Lorentz operation that approximates the generator $\{q^2, \circ\}$ [18], so as to have the paraxial axis-symmetric linear group $\mathrm{Sp}(2, \Re)$; we only have one of its solvable subgroups. In other words, we do not have yet[39] global "thin-lens" transformations corresponding even to the paraxial form $\mathbf{p} \mapsto \mathbf{p} + \alpha\mathbf{q}$.

Slightly over fifty years ago, Eugene P. Wigner published his classical paper [20] classifying the irreducible representations of the Poincaré group; possible free particles in Nature were identified as their bases, for each stratum of four-momenta and corresponding little group spins. Building on this, in a less well-known paper, François Lurçat [21] proposed, twenty-five years later, that fast decaying particles could be represented by poles on the complex plane of its coset manifolds, having a Breit-Wigner distribution of masses and spins. Marco Toller [22] took this picture to search for poles in the scattering cross section at forward momentum transfer to account geometrically for families of Regge poles that should provide a *kind* of dynamics for elementary particle scattering. Now, unlike the case in Schrödinger quantum mechanics, choosing interactions is notably difficult in field theories on the Poincaré group, and they remain basically as theories on empty space. We perceive the analogy with the Euclidean

[39] See the chapter by V.I. Man'ko and K.B. Wolf in this Volume, that addresses this problem and gives a solution.(Note added in proof).

group, in itself much simpler than the Poincaré group, in that homogeneous space rays and waves, polarized or scalar, are well described. But otherwise well-known elementary interactions such as refraction by a plane, already need their complex extension. One may hope that processes with a hidden symmetry, such as scattering by a refracting sphere, by a Maxwell fish-eye medium, say, or bound systems such as parabolic- or elliptic-profile fibers could be analyzed in terms analogous to hidden symmetry in quantum mechanics, keeping ray direction on its now complex Descartes sphere.

In Euclidean optics we *do* have transformations that globalize some of the second-order aberrations in the x and y–boosts, and for Seidel spherical aberration $\{(p^2)^2, \circ\}$ (in z–translations) and circular coma $\{p^2\mathbf{p}\cdot\mathbf{q}, \circ\}$ (in z–boosts). They have been constructed both in geometric and in wave optics, and generate an infinite dimensional solvable subgroup of all passive optical transformations. With this limitation and extent we answer the two questions posed in the first Section. Other issues that have been raised in the intervening material should be left for further development.

6.16 Acknowledgements

I would like to thank the participants of the Lie Methods in Optics II workshop for the generous hearing and comments on material from this contribution, when it was in embrionary form. In particular, I am indebted to Dr. R. Simon, who stayed on for a week at IIMAS–Cuernavaca, and provided the key remarks and silences for the structure to take form. Dr. Simon's visit to Mexico was made possible by the support from the **Third World Academy of Sciences** (Trieste) Fellowship program, and from the **Fondo de Fomento Educativo BCH** (México DF), as *Distinguished Visitor* of CIFMA AC. As I worked through the subject, I benefited from frequent consultations with Dr. François Leyvraz, Instituto de Física/Cuernavaca, that I wish to acknowledge.

6.17 References

[1] R. Hermann, *Vector Bundles in Mathematical Physics* (W.A. Benjamin, New York, 1970).

[2] R. Gilmore, *Lie groups, Lie Algebras, and Some of their Applications* (J. Wiley & Sons, New York, 1974).

[3] S. Steinberg, Lie series, Lie transformations, and their applications, in *Lie Methods in Optics*, Proceedings of the CIFMO–CIO Workshop (León, México, January 1985). Ed. by J. Sánchez-Mondragón and K.B. Wolf, Lecture Notes in Physics, Vol. 250 (Springer-Verlag, Heidelberg, 1986).

[4] K.B. Wolf and T.H. Seligman, Harmonic analysis on bilateral classes, *SIAM J. Math. Anal.* **11**, 1068–1074 (1980).

[5] H. Flanders, *Differential Forms with Applications to the Physical Sciences* (Academic Press, New York, 1963).

[6] K.B. Wolf, Symmetry in Lie Optics, *Ann. Phys.* **172**, 1–25 (1986).

[7] T. Sekiguchi and K.B. Wolf, The Hamiltonian formulation of optics, *Am. J. Phys.* **55**, 830–835 (1987).

[8] J. Durnin, Exact solutions for nondiffracting beams. I. The scalar theory, *J. Opt. Soc. Am.* **A4**, 651–654 (1987); J. Durnin, J.J. Miceli, and J.H. Eberly, Diffraction-free beams, *Phys. Rev. Lett.* **58**, 1499–1501 (1987).

[9] S. Steinberg and K.B. Wolf, Invariant inner products on spaces of solutions of the Klein-Gordon and Helmholtz equations, *J. Math. Phys.* **22**, 1660–1663 (1981).

[10] M.J. Bastiaans, *Local-Frequency Description of Optical Signals and Systems.* Eindhoven University of Technology Report 88-E-191 (April 1988). Lectures delivered at the First International School and Workshop in Photonics, Oaxtepec, June 28 – July 8, 1988.

[11] H. Raszillier and W. Schempp, Fourier optics from the perspective of the Heisenberg group. In *Lie Methods in Optics, op. cit.*

[12] K.B. Wolf, The Heisenberg–Weyl ring in quantum mechanics. In *Group Theory and its Applications, Vol. 3*, Ed. by E.M. Loebl (Academic Press, New York, 1975).

[13] J.-M. Lévy-Leblond, Galilei group and Galilean invariance. In *Group Theory and its Applications, Vol. 2*, Ed. by E.M. Loebl (Academic Press, New York, 1971).

[14] V. Guillemin and S. Sternberg, *Symplectic Techniques in Physics* (Cambridge University Press, 1984).

[15] C.P. Boyer and K.B. Wolf, Deformations of inhomogeneous classical Lie algebras to the algebras of the linear groups, *J. Math. Phys* **14**, 1853–1859 (1973); K.B. Wolf and C.P. Boyer,The algebra and group deformations $\mathrm{I}^m[\mathrm{SO}(n) \otimes \mathrm{SO}(m)] \Rightarrow \mathrm{SO}(n,m)$, $\mathrm{I}^m[\mathrm{U}(n) \otimes \mathrm{U}(m)] \Rightarrow \mathrm{U}(n,m)$, and $\mathrm{I}^m[\mathrm{Sp}(n) \otimes \mathrm{Sp}(m)] \Rightarrow \mathrm{Sp}(n,m)$, *J. Math. Phys* **15**, 2096–2101 (1974).

[16] N.M. Atakishiyev, W. Lassner, and K.B. Wolf, The relativistic coma aberration. I. Geometrical optics. Comunicaciones Técnicas IIMAS No. 509 (1988). To appear in *J. Math. Phys.*

[17] N.M. Atakishiyev, W. Lassner, and K.B. Wolf, The relativistic coma aberration. II. Helmholtz wave optics. Comunicaciones Técnicas IIMAS No. 517 (1988). To appear in *J. Math. Phys.*

[18] A.J. Dragt, E. Forest, and K.B. Wolf, Foundations of a Lie algebraic theory of geometrical optics. In *Lie Methods in Optics, op. cit.*

[19] M. Born and E. Wolf, *Principles of Optics* (Pergamon Press, Oxford, 1975), Chapter IX.

[20] E.P. Wigner, On unitary irreducible representations of the inhomogeneous Lorentz group, *Ann. Math.* **40**, 149–204 (1939). *See also* E.P. Wigner, Relativistic invariance and quantum phenomena, *Rev. Mod. Phys.* **29**, 255–268 (1957); D. Han, Y.S. Kim, and D. Son, Decomposition of Lorentz transformations, *J. Math. Phys.* **28**, 2373–2378 (1987).

[21] F. Lurçat, Quantum field theory and the dynamical role of spin, *Physics* **1**, 95–100 (1964). *cf.* G.N. Fleming, The spin spectrum of an unstable particle, *J. Math. Phys.* **13**, 626–637 (1974).

[22] M. Toller, Three-dimensional Lorentz group and harmonic analysis of the scattering amplitude, *N. Cimento* **37**, 631–657 (1965); *ibid.* An expansion of the scattering amplitude at vanishing four-momentum transfer using the representations of the Lorentz group, *N. Cimento* **53**, 671–715 (1967).

7

The map between Heisenberg-Weyl and Euclidean optics is comatic

Vladimir I. Man'ko and Kurt Bernardo Wolf

ABSTRACT The mathematics of coherent states is essentially a translation of oscillator quantum mechanics to the paraxial model of optics, and is based on the Heisenberg–Weyl algebra and group. On the other hand, "4π" optics is based on the three-dimensional Euclidean algebra and corresponding group. We show here that a global map between the two may be established. It is, in fact, third-order Seidel-Lie coma. Spherical and circular-comatic aberrations are a proper subgroup of the group of all canonical transformations of phase space, that can be subject to unique quantization and wavization.

7.1 Introduction

Our previous joint work [1] addressed the question of the behaviour of Gaussian beams, including Glauber coherent beams [2] and correlated coherent [4], [5] states, under spherical aberration. This is the only aberration present in free flight, and is increasingly important at large angles from the chosen optical axis [6].

The mathematical foundation of coherent state theory is the *Heisenberg-Weyl* Lie algebra of the quantum mechanical operators of position and momentum [7], [8], with the harmonic oscillator physical model of optical fibers of parabolic index profile [9]. This is a structure rich in results that has been profitably applied to laser optics, among many other fields. Recent work on *Euclidean* optics, *i.e.*, "4π" geometric and wave optics [10] motivated our enquiry into their exact relation. Physical three-dimensional optics requires $N = 2$ dimensional screens. We shall work in generic dimension N because, once we go beyond the simple $N = 1$ case, the mathematical formulation allows easily such generalization.

It is generally taken for granted that Heisenberg-Weyl optics is the *paraxial limit*, meaning small angles and distances from the optical axis, of 4π optics. We believe that Heisenberg-Weyl and Euclidean optics should be described as *separate* mathematical structures. We show that there exists a *semi*-global 1:1 map between the two, that assigns to each point in a $\Re^{2N}$

Heisenberg-Weyl phase space one point in the Euclidean phase space of $2N$ dimensions where the optical momentum values are bounded and rays are in the *forward* hemisphere, *i.e.*, advance in the direction of the optical axis. Another $\Re^{2N}$ space maps on the *backward* hemisphere, and continuity conditions are asked to hold between the former two 'flat' spaces.

The map between the Heisenberg-Weyl and Euclidean optics is a *point* Escher-like transformation between the N-dimensional momentum space of the former and the interior of the N-dimensional momentum sphere of the latter. In the $N = 2$ case of ordinary optics, the latter sphere interior is a *disk* of radius n, the refractive index of the medium. The canonically conjugate observables of ray *position* are then related by a map that is in general *circular comatic*, and in fact *precisely* third-order Lie-Seidel coma. This intertwining of Heisenberg-Weyl and Euclidean optics allows the linear (and nonlinear) symplectic transformations of the former to be applied to the latter. And conversely, rotations [11] and relativistic boosts [12], that act naturally on the sphere [10], are thereby applied on the $\Re^N$ space of Heisenberg-Weyl optics. The map is near to unity in the paraxial region, *i.e.*, small angles and distances from the optical axis, but extends analytically to all angles and distances. This we can do for both geometric and wave optics.

Section 2 succintly derives the Hamilton equations of motion for optics directly from Snell's law in local form. The proof of this was presented in Ref. [13] by a didactical route that is considerably shorter than that of Fermat's principle through the Lagrangian formalism [6]. The present approach is adapted from Ref. [14]. The paraxial approximation is treated in Section 3, and is extended to the full Heisenberg-Weyl phase space. We compare the (straight) trajectories of rays freely propagating in this and in Euclidean optics and introduce, in Section 4, the *opening coma* transformation between them. Section 5 then shows that in this way, indeed, we intertwine the two different régimes of propagation in homogeneous optical media.

Heisenberg-Weyl optics has a distinguished transformation group, that is *larger* than the Heisenberg-Weyl group itself [7]: *linear* symplectic transformations. Euclidean optics, on the other hand, accomodates naturally Euclidean and Lorentz transformations. In Sections 6 and 7 we apply the former group to the latter phase space, and *viceversa*. Such cross-applications constitute apparently novel nonlinear realizations of these groups. Section 8 is set towards structuring the results for general spherical plus comatic aberration maps of phase space.

The opening coma map in *wave* optics is seen from Section 9 on. From the $\mathcal{L}_2(\Re^N)$ inner product, through the map, we arrive at the Hilbert space $\mathcal{H}_k^N$ of oscillatory solutions of the Helmholtz equation with an inner product that is nonlocal. In Sections 10 and 11 we follow through the coma map the plane waves, δ's, Gaussians (including coherent and correlated coherent states), and Bessel-function nondiffracting beams. They are analogous but distinct in the paraxial (Heisenberg-Weyl) and in the global (Euclidean) regimes. Section 11 ends with the necessary context to present some open directions into fields where the formalism of Heisenberg and Weyl has been successful and where it could be extended to all ray directions.

7.2 From Snell's law to the Hamilton equations

We consider a homogeneous optical medium (generally, of $N+1$ dimensions), characterized by a refractive index n, separated by a surface σ from a second such medium of different index n'. We assume σ has a tangent plane $T_\sigma(S)$ at each point $S \in \sigma$, characterized by its normal $(N+1)$-dimensional vector $\vec{\Sigma}(S)$. Similarly, we denote by $\vec{n}$ and $\vec{n}'$ the ray direction vectors that range over spheres of radii given by $|\vec{n}| = n$ and $|\vec{n}'| = n'$, the two refractive indices. These are the *Descartes* spheres of rays in the two media. The 'rays' may be one-dimensional straight lines in space, or may be N dimensional *planes* with normals $\vec{n}$ and $\vec{n}'$ [10].

Snell's law is the statement that the change in the direction vector $\vec{n}-\vec{n}'$ be *along* the refracting surface normal $\vec{\Sigma}$ *at* the incidence point of the ray. When $n = n'$, then also $\vec{n} = \vec{n}'$.

In ordinary optics ($N+1=3$), the law may be cast in the cross-product form $\vec{\Sigma} \times \vec{n} = \vec{\Sigma} \times \vec{n}'$. The equality of the norm of this relation implies the familiar sine law $n \sin\theta = n' \sin\theta'$, with θ being the angle between $\vec{\Sigma}$ and $\vec{n}$, and similarly for the primed. This relation also tells us that the three vectors are in a plane. Equally important, but generally left unsaid, is that the incidence point S of the ray in medium n is the *same* as its departure point S in medium n'.

Snell's law in $N+1$ dimensions implies that the projection of $\vec{n}$ on the tangent plane $T_\sigma(S)$ is *conserved.* This is the N-dimensional vector of optical *momentum* referred to the tangent plane of the refracting surface. The N coordinates of the incidence point S on the screen are also conserved.

Let us choose a standard Cartesian coordinate system in space where points are $(N+1)$-vectors $\vec{q}$, where rays are *geometrical* lines $\vec{q}(z) = (q_1(z), q_2(z), \ldots, q_N(z), z) = (\mathbf{q}(z), z)$, $z \in \Re$ being the $(N+1)^{\text{th}}$ coordinate taken as the *optical axis*, indicating by **boldface** the first N components $\mathbf{q}(z)$, the *position* of the ray at the *standard screen:* the $z=0$ plane . This parametrization fails only for rays paralell to the screen; it is not a singularity of the space of rays but only indicates the limit of this particular coordinate chart.

The ray direction $\vec{n}$ is the $(N+1)$-dimensional vector tangent to $\vec{q}(z)$, and will depend on z when the medium is inhomogeneous through $n(\vec{q})$. The Cartesian components of $\vec{n}$ are not all independent since this vector lies on the surface of the Descartes sphere of radius n. It is convenient to single out again the $(N+1)^{\text{th}}$ component and write $\vec{n} = (p_1, p_2, \ldots, p_N, h) = (\mathbf{p}, h)$, where

$$h = \tau\sqrt{n^2 - p^2}, \qquad p^2 = \mathbf{p}\cdot\mathbf{p}, \quad \tau \in \{+, 0, -\}\ (\tau = 0 \text{ when } h = 0). \quad (2.1)$$

The vector $\vec{n} = (\mathbf{p}, h)$, tangent to $\vec{q}(z)$, is thus paralell to $d\vec{q}(z) = (d\mathbf{q}, dz)$. Hence, the following proportion holds between each of their first N components,

$$\frac{d\mathbf{q}}{dz} = \frac{\mathbf{p}}{h} = -\frac{\partial h}{\partial \mathbf{p}}. \qquad (2.2)$$

The second equality is a consequence of the specific function form of h on $\mathbf{p}$ given in equation (2.1).[1] The equality between the first and last members in (2.2) is the *first* Hamilton equation. The origin of this equation is thus purely geometrical. If the medium is inhomogeneous through $n(\vec{q})$, infinitesimal changes in ray direction contribute only to second differentials in $\mathbf{q}$, so they do not appear in the first-order Hamilton equation.[2]

The *second* Hamilton equations are *dynamical* and speak of the inhomogeneity of the medium $n(\mathbf{q}, z)$ through the obeyance of $\vec{n}(z)$ to Snell's law. Let the gradient of the refractive index $\vec{\nabla} n = (\partial n/\partial \mathbf{q}, \partial n/\partial z)$ take the role of the surface normal $\vec{\Sigma}$ above. This vector will be parallel to the *change* of the direction vector, $d\vec{n}/dz = (d\mathbf{p}/dz, dh/dz)$. Let the ratio be a certain $\alpha(\mathbf{p}, \mathbf{q}, z)$ that we shall find through the constraint that the direction vector remain on its Descartes sphere, $\vec{n}\cdot\vec{n} = n^2 = p^2 + h^2$.

Thus, on the one hand,

$$\frac{d}{dz}n^2 = \frac{d}{dz}\vec{n}\cdot\vec{n} = 2\vec{n}\cdot\frac{d\vec{n}}{dz} = 2\alpha\vec{n}\cdot\vec{\nabla} n, \qquad (2.3a)$$

and on the other, using (2.2) and the chain rule,

$$\begin{aligned}\frac{d}{dz}n^2 &= 2n\left(\frac{d\mathbf{q}}{dz}\cdot\frac{\partial n}{\partial \mathbf{q}} + \frac{\partial n}{\partial z}\right) = 2n\left(\frac{\mathbf{p}}{h}\cdot\frac{\partial n}{\partial \mathbf{q}} + \frac{\partial n}{\partial z}\right) \\ &= 2\frac{n}{h}\left(\mathbf{p}\cdot\frac{\partial n}{\partial \mathbf{q}} + h\frac{\partial n}{\partial z}\right) = 2\frac{n}{h}\vec{n}\cdot\vec{\nabla} n. \qquad (2.3b)\end{aligned}$$

From the equality of the last members we conclude that $\alpha = n/h$. Consequently,

$$\frac{d\vec{n}}{dz} = \frac{n}{h}\vec{\nabla} n. \qquad (2.4)$$

Using again the function form of h, we write in components

$$\frac{d\mathbf{p}}{dz} = \frac{n}{h}\frac{\partial n}{\partial \mathbf{q}} = \frac{\partial h}{\partial \mathbf{q}}, \qquad (2.5a)$$

---

[1] We use the vector derivative notation $\partial/\partial\mathbf{p} = (\partial/\partial p_1, \partial/\partial p_2, \ldots, \partial/\partial p_N)$.

[2] Second differentials help to build the *ray* differential equation [15], that is of second order.

$$\frac{dh}{dz} = \frac{n}{h}\frac{\partial n}{\partial z} = \frac{\partial h}{\partial z}. \qquad (2.5b)$$

The first and last members of these equalities constitute de *second* Hamilton equations of motion.

In the case of homogeneous media, $n =$ constant, the equations of motion (2.5) and (2.2) become, respectively,

$$\frac{d\mathbf{p}}{dz} = 0, \qquad \frac{d\mathbf{q}}{dz} = \frac{\mathbf{p}}{h} = \frac{\mathbf{p}}{\tau\sqrt{n^2 - p^2}}. \qquad (2.6a, b)$$

The solution to these equations in terms of initial ($z = 0$) *screen* values $\mathbf{p}_0$ and $\mathbf{q}_0$, is the ray path

$$\mathbf{p}(z) = \mathbf{p}_0, \qquad \mathbf{q}(z) = \mathbf{q}_0 + z\frac{\mathbf{p}_0}{\tau\sqrt{n^2 - p_0^2}}. \qquad (2.7a, b)$$

This is a z-dependent *canonical* transformation [16] that conserves the Poisson brackets

$$\{q_i(z), p_k(z)\} = \delta_{i,k}, \qquad i, k = 1, 2, \ldots, N. \qquad (2.8a)$$

$$\{q_i(z), q_k(z)\} = \{p_i(z), p_k(z)\} = 0. \qquad (2.8b)$$

The validity of the Hamilton equations (2.2) and (2.5) establishes $\mathbf{q}$ and $\mathbf{p}$ as *canonically conjugate* quantities of optical position and momentum. Position $\mathbf{q} \in \Re^N$ is the manifold of the standard screen plane; optical momentum $\mathbf{p}$ ranges over the interior of the N-sphere $|\mathbf{p}| < n$ (a *disk* in $N = 2$, 3-dimensional ordinary optics) once for forward rays ($h > 0$, $\tau = +$), and once for backward rays ($h < 0$, $\tau = -$). The two charts are separated by rays in the equator of the Descartes sphere ($h = 0$, $\tau = 0$), where we expect continuity conditions to hold. We shall speak for the most part of *forward* rays $\tau = +$.

Evolution along the z-axis of Euclidean geometrical optics is thus ruled by the Hamiltonian function

$$\begin{aligned} H^E &= -h = -\sqrt{n^2 - p^2} \\ &= -n + \frac{p^2}{2n} + \frac{(p^2)^2}{8n^3} + \frac{(p^2)^3}{16n^5} + \cdots, \end{aligned} \qquad (2.9)$$

that is (minus) the $(N+1)^{\text{th}}$-component of the ray direction vector. The series expansion will be now subject to scrutiny.

7.3 The paraxial régime and Heisenberg–Weyl optics

We consider now the *paraxial* optical régime, *i.e.*, the approximation that is valid when the angle θ between rays $\vec{n}$ in a light beam and the chosen optical z-axis is small. In (hyper) spherical coordinates, this angle is given by the rays' momentum $\mathbf{p}$ through

$$|\mathbf{p}| = n \sin\theta. \tag{3.1a}$$

The paraxial approximation thus entails selecting a region of optical phase space restricted by

$$|\theta| \ll \pi, \quad i.e., \quad |\mathbf{p}| \ll n. \tag{3.1b}$$

For such beams the Hamiltonian (2.9), expanded in power series with respect to p^2/n^2, is approximated by the *paraxial* form

$$H^E \approx \frac{p^2}{2n} - n. \tag{3.2}$$

If the refractive index n is a constant, this is the Hamiltonian for mechanical *free motion* in N-dimensional space. The corresponding phase space is $2N$-dimensional, and its basic group of motions is the Heisenberg-Weyl group. Let us consider now *this* Hamiltonian and *its* free motion dynamics. We may omit the constant term $-n$ in (3.2) since its Lie operator is zero, and describe the free motion system by the Heisenberg-Weyl Hamiltonian

$$H^{HW} = \frac{P^2}{2n}, \qquad P^2 = \mathbf{P}\cdot\mathbf{P}. \tag{3.3}$$

Throughout this paper, we shall use capital letters $\mathbf{P}$ and $\mathbf{Q}$ to indicate that the observables belong to the Heisenberg-Weyl phase space while lower-case $\mathbf{p}$ and $\mathbf{q}$ will remain for Euclidean optical variables.

The Hamilton equations of motion in phase space (2.2)–(2.5) for the free point particle are

$$\frac{d\mathbf{P}}{dz} = -\frac{\partial H^{HW}}{\partial \mathbf{Q}} = 0, \qquad \frac{d\mathbf{Q}}{dz} = \frac{\partial H^{HW}}{\partial \mathbf{P}} = \frac{\mathbf{P}}{n}. \tag{3.4a, b}$$

The solution to these equations is

$$\mathbf{P}(z) = \mathbf{P}_0, \qquad \mathbf{Q}(z) = \mathbf{Q}_0 + z\mathbf{P}_0/n, \tag{3.5a, b}$$

where $\mathbf{P}_0$ and $\mathbf{Q}_0$ are the initial momenta and positions in phase space. Compare this with equations (2.6) and solutions (2.7) in Euclidean optics, which are also straight lines in space. These relations are here *linear*, and may be written in matrix form through $2N \times 2N$ *symplectic* matrices with

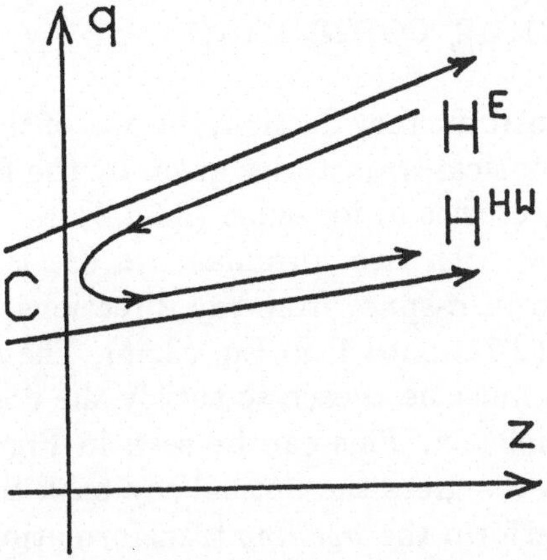

FIGURE 1. The map from Euclidean to Heisenberg-Weyl propagation: The ray coordinates are regressed back to the screen $z = 0$; next, the *opening* of Euclidean to Heisenberg-Weyl ranges; lastly, we evolve the ray from the screen to general z in the Heisenberg-Weyl regime, *i.e.*, paraxial optics.

$N \times N$ blocks that are proportional to the identity matrix. Such transformations are also canonical since they preserve the Poisson bracket as in (2.8).

What are the phase space *bounds* of this paraxial construction? In principle none that may be naturally incorporated in the Heisenberg-Weyl algebra with basic Poisson brackets (2.8) or the corresponding Lie *groups*.[3] Moreover, quantum mechanics (in the standard theory) *demands* that observables have self-adjoint operators to their name, and the familiar Schrödinger or coherent-state generators of the Heisenberg-Weyl algebra need the full real line, both for $\mathbf{Q}$ and $\mathbf{P}$. Finally, the possibility of using the linear symplectic group for paraxial optics also hinges on the assumption that the range of momentum $\mathbf{P}$ be the full plane $\Re^N$. In fact, we *know* that nonrelativistic mechanics is a consistent mathematical theory of much practical use in paraxial optics, and the basis for aberration expansions into the metaxial regime. But we stress that it is a theory globally *different* from the 4π Euclidean optics presented in last Section. We shall now proceed to find the map between the systems obeying H^E and H^{HW}.

[3]In reference [17] we examined a 1-dimensional Heisenberg-Weyl group where the momentum parameter is cyclic —direction lies on a circle in two-dimensional optics. This entailed that the conjugate position parameter be *discrete* (*cf.* the sampling theorem) denying thus the possiblity of having an infinitesimal translation generator. The central subgroup is also forced to be cyclic —as it *should*, being a phase. The N-dimensional version of that construction yields the direction vector as ranging over an N–torus, however, instead of an N–sphere.

7.4 The opening coma map

As we stated in the introductory Section, the aim of this work is to construct the map of the free optical trajectories given by the formulae (2.7a, b) onto the free motion trajectories in formulae (3.5a, b).

In fact, comparing each two formulae, we see that we must find the point transformation in **p**-space from ray directions, that maps the factor $\mathbf{p}/\sqrt{n^2 - p^2}$ in Eq. (2.7b) onto **P** in Eq. (3.5b). The canonically conjugate position coordinates must be chosen to satisfy the demand of *canonicity* of the desired transformation. This can be seen in Figure 1, and consists of three factors: first, we regress the optical ray back to the standard screen $z = 0$; second, we perform the *opening* transformation from the Euclidean to the Heisenberg-Weyl variables, and in such a way that the two different momentum ranges are properly related; and third, we evolve the ray from the screen to general z in the Heisenberg-Weyl regime.

The first factor in the map will be thus the transformation *inverse* to (2.7), *i.e.*, *backward* free propagation of Euclidean rays; this we may write in terms of a Lie transformation [16]: [4]

$$\mathbf{p}_0 = \mathbf{p}(0) = \exp(+z\hat{H}^{E})\mathbf{p}(z) = \mathbf{p}(z), \tag{4.1a}$$

$$\mathbf{q}_0 = \mathbf{q}(0) = \exp(+z\hat{H}^{E})\mathbf{q}(z) = \mathbf{q}(z) - z\frac{\mathbf{p}(z)}{\sqrt{n^2 - \mathbf{p}(z)^2}}. \tag{4.1b}$$

The last transformation, Eqs. (3.5), is the *forward* evolution in Heisenberg-Weyl mechanical space. It may be similarly expressed as a Lie exponential

$$\mathbf{P}(z) = \exp(-z\hat{H}^{HW})\mathbf{P} = \mathbf{P}(0), \tag{4.2a}$$

$$\mathbf{Q}(z) = \exp(-z\hat{H}^{HW})\mathbf{Q} = \mathbf{Q}(0) + z\mathbf{P}(0)/n. \tag{4.2b}$$

[4] We recall that, associated to every differentiable function f, we define its Lie *operator*

$$\hat{f} = \{f, \circ\} = \sum_{i=1}^{n} \left(\frac{\partial f}{\partial q_i}\frac{\partial}{\partial p_i} - \frac{\partial f}{\partial p_i}\frac{\partial}{\partial q_i} \right).$$

These operators have the important property of intertwining with commutators,

$$\{f, g\}^{\wedge} = [\hat{f}, \hat{g}],$$

so that we may speak of the Lie *algebra* generated by a set of functions under the Poisson bracket. The Lie *transformation* generated by f is the exponential of its Lie operator,

$$\exp \hat{f} = \sum_{n=1}^{\infty} \frac{1}{n!}(\hat{f})^n = \sum_{n=1}^{\infty} \frac{1}{n!}\{f, \{f, \cdots \{f, \circ\} \cdots\}\}.$$

These transformations are well known to be canonical.

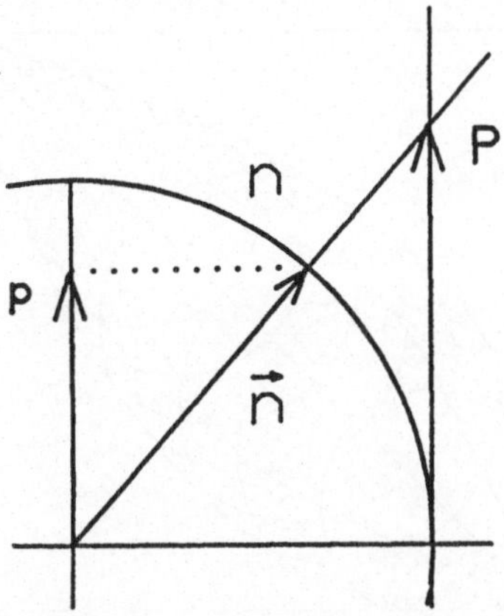

FIGURE 2. The geometric map from Euclidean to Heisenberg-Weyl momenta, **p** in the former and **P** in the latter, $\vec{n}$ is the ray direction vector.

The basic *opening* transformation at the screen is the following:

$$\mathbf{p} \mapsto \mathbf{P} \;=\; \mathcal{C}\mathbf{p} = \frac{\mathbf{p}}{\sqrt{1-p^2/n^2}}, \tag{4.3c}$$

$$\mathbf{q} \mapsto \mathbf{Q} \;=\; \mathcal{C}\mathbf{q} = \sqrt{1-p^2/n^2}\left(\mathbf{q} - \frac{1}{n^2}\mathbf{p}\cdot\mathbf{q}\,\mathbf{p}\right). \tag{4.3d}$$

This transformation *opens* the compact Euclidean momentum range $|\mathbf{p}| < n$ to the full Heisenberg-Weyl momentum plane $\Re^N$. We have called this map *comatic* because it is in fact a particular case of a Lie transformation [18], generated by the coma monomial ${}^2\mathcal{X}_1^2 = p^2\mathbf{p}\cdot\mathbf{q}$ [14], [19]

$$\mathbf{P}_\gamma \;=\; \exp(\gamma p^2\mathbf{p}\cdot\mathbf{q})\hat{}\,\mathbf{p} = \mathbf{p}/\sqrt{1-2\gamma p^2} \tag{4.4a}$$

$$\;=\; \mathbf{p} + \gamma p^2\mathbf{p} + \tfrac{3}{2}\gamma^2(p^2)^2\mathbf{p} + \tfrac{5}{2}\gamma^3(p^2)^3\mathbf{p} + \cdots. \tag{4.4b}$$

$$\mathbf{Q}_\gamma \;=\; \exp(\gamma p^2\mathbf{p}\cdot\mathbf{q})\hat{}\,\mathbf{q} = \sqrt{1-2\gamma p^2}(\mathbf{q} - 2\gamma\mathbf{p}\cdot\mathbf{q}\,\mathbf{p}) \tag{4.4c}$$

$$\begin{aligned} &= \mathbf{q} - \gamma(p^2\mathbf{q} - 2\mathbf{p}\cdot\mathbf{q}\,\mathbf{p}) \\ &\quad - \gamma^2p^2(\tfrac{1}{2}p^2\mathbf{q} - 2\mathbf{p}\cdot\mathbf{q}\,\mathbf{p}) \\ &\quad - \gamma^3(p^2)^2(\tfrac{1}{2}p^2\mathbf{q} - \mathbf{p}\cdot\mathbf{q}\,\mathbf{p}) - \cdots, \end{aligned} \tag{4.4d}$$

for the value $\gamma = 1/2n^2$ of the parameter. To third degree in the phase space variables we have the usual third-order Seidel coma of optics; the full Lie series defines the *Seidel-Lie* global coma aberration.

In Figure 2 we show the geometric map between the Heisenberg-Weyl momentum plane and the 'forward' half-circle of ray directions and its projection on the Euclidean momentum segment $(-n, n)$. Correspondingly, in Figure 3 we show the position variables $\mathbf{q}'(\mathbf{p}, \mathbf{q})$ (in $N = 2$ dimensions)

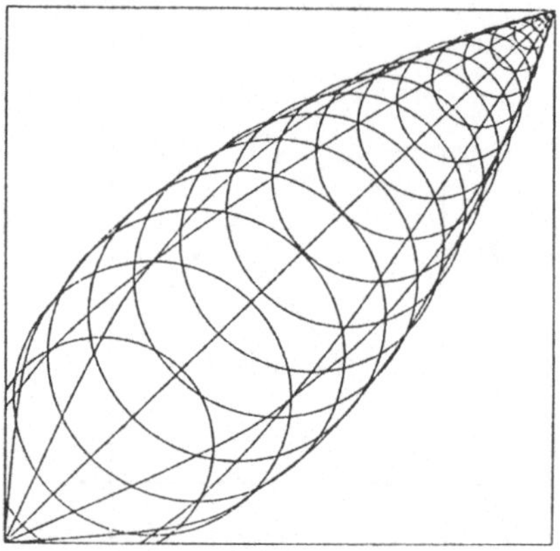

FIGURE 3. The spot diagram for the 'system' $\mathcal{C}$. A 90° pencil of rays in 5°-intervals in paralells and 16 meridians, issuing from a point at the upper right corner, upon passage through this '$\mathcal{C}$-system', maps on the indicated points on the screen.[5]

for $\mathbf{q}$ fixed and letting $\mathbf{p}$ range over a polar coordinate grid around the optical axis, *i.e.*, the *spot diagram* of the 'system' $\mathcal{C}$.

The vector function $A(p^2)p^2\mathbf{q} + B(p^2)\mathbf{p}\cdot\mathbf{q}\,\mathbf{p}$ maps a cone of rays ($p^2 =$ constant) twice onto a *circle* on the screen, of radius $\frac{1}{2}Bp^2q$, $q = |\mathbf{q}|$, and center at $(A+\frac{1}{2}B)p^2q$. These circles are tangent to a sector with apex at $\mathbf{q}$ and half-angle of 30° at the apex. Figure 3 shows that as $|\mathbf{p}|$ increases from 0 to n, the radii grow from zero to a maximum and then decrease back to zero. The segment of circle *centers* starts at $\mathbf{q}$ and ends at the optical center $\mathbf{q} = 0$. This is the *global spot diagram* of the coma transformation.

The *inverse* comatic map $\mathcal{C}^{-1}$ may be obtained from (4.4)–(4.5) for the negative value of the parameter $\gamma = -1/2n^2$. This yields the inverse of (4.3) to be

$$\mathbf{p} = \mathcal{C}^{-1}\mathbf{P} = \frac{\mathbf{P}}{\sqrt{1+P^2/n^2}}, \qquad (4.6a)$$

$$\mathbf{q} = \mathcal{C}^{-1}\mathbf{Q} = \sqrt{1+P^2/n^2}\left(\mathbf{Q} + \frac{1}{n^2}\mathbf{P}\cdot\mathbf{Q}\,\mathbf{P}\right). \qquad (4.6b)$$

This map *closes* the noncompact range of Heisenberg-Weyl 'directions' into the compact region of Euclidean ray directions. Figure 4 shows the spot diagram of the 'system' $\mathcal{C}^{-1}$; the comet-like sectors in the spot diagram are now unbounded. We note the following useful identities under the

[5]Figures 3 and 4 have been prepared using the program SPOT_D developed by Guillermo Correa at IIMAS-UNAM. See: G.J. Correa-Gómez and K.B. Wolf, SPOT_D, *Programa para graficación de Diagramas de Manchas en Optica.* Comunicaciones Técnicas IIMAS, Serie Desarrollo, No. 97 (1989).

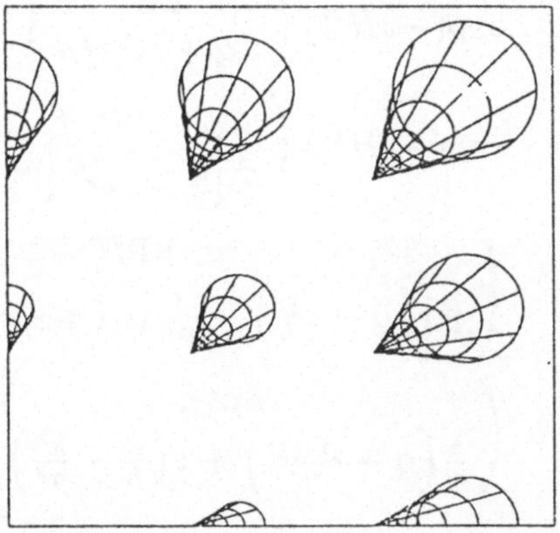

FIGURE 4. The spots diagram for the 'system' $\mathcal{C}^{-1}$. Rays with Heisenberg-Weyl $|\mathbf{P}| \leq n/2$ (corresponding to $P = n \sin \Theta$, $|\Theta| \leq 30°$), in six intervals of 5°, issuing from a rectangular grid of points at the apeces of the comet (*coma*) spot.

transformation:

$$1 - p^2/n^2 = \frac{1}{1 + P^2/n^2}, \quad 1 + P^2/n^2 = \frac{1}{1 - p^2/n^2}, \tag{4.7a}$$

$$\mathbf{p} \cdot \mathbf{q} = \mathbf{P} \cdot \mathbf{Q}(1 + P^2/n^2), \quad \mathbf{P} \cdot \mathbf{Q} = \mathbf{p} \cdot \mathbf{q}(1 - p^2/n^2). \tag{4.7b}$$

7.5 The map between Heisenberg-Weyl and Euclidean free rays

We have the backward transformation to the reference screen $z = 0$ for Euclidean rays in (4.1), the basic comatic map at that reference screen in (4.3), and the forward evolution from the screen to general z in (4.2). We now compose the three Lie transformations in that order:[6]

$$\begin{pmatrix} \mathbf{P}(z) \\ \mathbf{Q}(z) \end{pmatrix} = \exp(-zH^E)\, \mathcal{C}\, \exp(zH^{HW}) \begin{pmatrix} \mathbf{p}(z) \\ \mathbf{q}(z) \end{pmatrix}. \tag{5.1}$$

To compose the Lie transformations we consider the three operators acting on *dummy* variables that are called simply $\mathbf{p}$ and $\mathbf{q}$. We do this explicitly:

$$\begin{pmatrix} \mathbf{P}(z) \\ \mathbf{Q}(z) \end{pmatrix} = \exp(-zH^E)\, \mathcal{C} \begin{pmatrix} \mathbf{p} \\ \mathbf{q} + z\mathbf{p}/n \end{pmatrix} \tag{5.2a}$$

[6] We recall that we use capital $\mathbf{P}$, $\mathbf{Q}$ for Heisenberg-Weyl phase space variables of N-dimensional mechanical systems, and lower-case $\mathbf{p}$, $\mathbf{q}$ for the N-dimensional screen in $(N+1)$-dimensional space.

$$= \exp(-zH^E)\begin{pmatrix} \mathcal{C}\mathbf{p} \\ \mathcal{C}\mathbf{q} + z\mathcal{C}\mathbf{p}/n \end{pmatrix} \tag{5.2b}$$

$$= \exp(-zH^E)\begin{pmatrix} n\mathbf{p}/h \\ \frac{h}{n}\left[\mathbf{q} - \frac{\mathbf{p}\cdot\mathbf{q}\mathbf{p}}{n^2}\right] + z\frac{\mathbf{p}}{h} \end{pmatrix} \tag{5.2c}$$

$$= \begin{pmatrix} n\mathbf{p}/h \\ \frac{h}{n}\left[\left(\mathbf{q} - z\frac{\mathbf{p}}{h}\right) - \frac{1}{n^2}\mathbf{p}\cdot\left(\mathbf{q} - z\frac{\mathbf{p}}{h}\right)\mathbf{p}\right] + z\frac{\mathbf{p}}{h} \end{pmatrix} \tag{5.2d}$$

$$= \begin{pmatrix} n\mathbf{p}/h \\ \frac{h}{n}\left(\mathbf{q} - \frac{\mathbf{p}\cdot\mathbf{q}\mathbf{p}}{n^2}\right) + z\left(\frac{n}{h} - \frac{h^2}{n^2}\right)\frac{\mathbf{p}}{n} \end{pmatrix}, \tag{5.2e}$$

where as before we have abbreviated $h = \tau\sqrt{n^2 - p^2}$, and consider the $\tau = +$ hemisphere of rays.

The z-dependent map that relates the (lower case) Euclidean and (upper-case) Heisenberg-Weyl trajectories is thus

$$\mathbf{P} = \frac{\mathbf{p}}{\sqrt{1 - p^2/n^2}}, \qquad \text{independent of } z, \tag{5.3a}$$

$$\mathbf{Q}(z) = \sqrt{1 - p^2/n^2}\left(\mathbf{q}(z) - \frac{\mathbf{p}\cdot\mathbf{q}(z)\,\mathbf{p}}{n^2}\right) + z\left(\frac{1}{\sqrt{1 - p^2/n^2}} - \left[1 - \frac{p^2}{n^2}\right]\right)\frac{\mathbf{p}}{n}. \tag{5.3b}$$

At $z = 0$ we have the simple comatic map $\mathcal{C}$ in (4.3). In the process of z-evolution, the maps start to differ from the pure comatic one by a term of *spherical aberration* that increases linearly with z. Finally, the map *inverse* to (5.3) is

$$\mathbf{p} = \frac{\mathbf{P}}{\sqrt{1 + P^2/n^2}}, \qquad \text{independent of } z, \tag{5.4a}$$

$$\mathbf{q}(z) = \sqrt{1 + P^2/n^2}\left(\mathbf{Q}(z) + \frac{\mathbf{P}\cdot\mathbf{Q}(z)\,\mathbf{P}}{n^2}\right) + z\left(\frac{1}{\sqrt{1 + P^2/n^2}} - \left[1 + \frac{P^2}{n^2}\right]\right)\frac{\mathbf{P}}{n}. \tag{5.4b}$$

We could be tempted to think that the map between trajectories may be obtained dispensing with the middle transformation $\mathcal{C}$, simply regressing the Euclidean trajectory to the screen and taking this as the initial ray parameters to advance under the Heisenberg-Weyl free-flight Hamiltonian. We must remember, however, that the two phase spaces are essentially *different* and the $\mathcal{C}$ and $\mathcal{C}^{-1}$ maps are therefore needed to intertwine the two ranges for the momentum variables, $|\mathbf{p}| < n$ and $\mathbf{P} \in \Re^N$.

7.6 The symplectic group on Euclidean phase space

The map between Heisenberg-Weyl and Euclidean phase spaces allows us to find explicitly the action of the inhomogeneous linear symplectic group $\mathrm{ISp}(2N, \Re)$, well known in its Heisenberg-Weyl realization, on the N-sphere interior $|\mathbf{p}| < n$, the true phase space of optics. For the case of ordinary optics ($N = 2$ screen dimensions) the group is the four-dimensional inhomogeneous symplectic group $\mathrm{ISp}(4, \Re)$.

The $2N + 1$ functions whose Lie operators generate the N-dimensional Heisenberg-Weyl group of phase space translations and its center are

$$P_i = \frac{p_i}{\sqrt{1 - p^2/n^2}}, \quad Q_j = \sqrt{1 - p^2/n^2}\left(q_j - \frac{\mathbf{p}\cdot\mathbf{q}\,p_j}{n^2}\right), \quad 1, \qquad (6.1a, b, c)$$

for $i, j = 1, 2, \ldots, N$. The $\frac{1}{2}N(N + 1)$ generators of the linear transformations are

$$\begin{aligned} P_i P_j &= p_i p_j/(1 - p^2/n^2), && (6.2a)\\ P_i Q_j &= p_i q_j + \mathbf{p}\cdot\mathbf{q}\,p_i p_j/n^2, && (6.2b)\\ Q_i Q_j &= (1 - p^2/n^2) \\ &\quad \times\left(q_i q_j - \mathbf{p}\cdot\mathbf{q}(q_i p_j + q_j p_i)/n^2 + (\mathbf{p}\cdot\mathbf{q})^2 p_i p_j/n^4\right). && (6.2c) \end{aligned}$$

These functions close into the algebra of $\mathrm{ISp}(2N, \Re)$ under the Poisson bracket. The maximal compact subgroup of $\mathrm{Sp}(2N, \Re)$ is the rotation subgroup $\mathrm{SO}(N)$ within the N-dimensional screen generated by the linear combinations $R_{i,j} = P_i Q_j - P_j Q_i = p_i q_j - p_j q_i$. These generators also belong to the Euclidean algebra and will be seen below. The part of $\mathrm{ISp}(2N, \Re)$ that has zero Lie brackets with the $R_{i,j}$ is the subalgebra $\mathrm{Sp}(2, \Re)$ of *axis-symmetric* systems. Explicitly, the generators of this 'radial' $\mathrm{Sp}(2, \Re)$ are

$$\begin{aligned} P^2 &= p^2/(1 - p^2/n^2), && (6.3a)\\ \mathbf{P}\cdot\mathbf{Q} &= (1 - p^2/n^2)\,\mathbf{p}\cdot\mathbf{q}, && (6.3b)\\ Q^2 &= (1 - p^2/n^2)\left(q^2 - 2(\mathbf{p}\cdot\mathbf{q})^2/n^2 + p^2(\mathbf{p}\cdot\mathbf{q})^2/n^4\right). && (6.3c) \end{aligned}$$

By exponentiation, (6.1) generate the finite translations

$$\exp \mathbf{E}\cdot\hat{\mathbf{P}}\begin{pmatrix}\mathbf{P}\\ \mathbf{Q}\end{pmatrix} = \begin{pmatrix}\mathbf{P}\\ \mathbf{Q} - \mathbf{E}\end{pmatrix}, \quad \exp \mathbf{F}\cdot\hat{\mathbf{Q}}\begin{pmatrix}\mathbf{P}\\ \mathbf{Q}\end{pmatrix} = \begin{pmatrix}\mathbf{P} + \mathbf{F}\\ \mathbf{Q}\end{pmatrix}. \qquad (6.4a, b)$$

The quadratic functions (6.2) generate *linear* transformations that we may put in matrix form as

$$\begin{pmatrix}\mathbf{P}\\ \mathbf{Q}\end{pmatrix} \overset{\mathcal{S}}{\mapsto} \mathbf{S}\begin{pmatrix}\mathbf{P}\\ \mathbf{Q}\end{pmatrix}, \qquad \mathbf{S} = \begin{pmatrix}\mathbf{A} & \mathbf{B}\\ \mathbf{C} & \mathbf{D}\end{pmatrix}, \qquad (6.5a)$$

where $\mathbf{S}$ is a *symplectic* matrix, *i.e.*, that satisfies

$$\mathbf{S}\mathbf{M}\mathbf{S}^{\top} = \mathbf{M}, \qquad \mathbf{M} = \begin{pmatrix} 0 & 1 \\ -1 & 0 \end{pmatrix}, \tag{6.5b}$$

where $^{\top}$ indicates transposition, and $\mathbf{M}$ is the symplectic *metric* matrix. The 2×2 block realization for $\mathbf{S}$ leads to well-known relations between the $N \times N$ block matrices [6]. The axis-symmetric linear combinations (6.3) generate the subset of matrices (6.3*b*) where the submatrices are multiples of the $N \times N$ unit matrix.

The action of the Heisenberg-Weyl group on Euclidean phase space coordinates, for "position"-translations (6.4*a*), is

$$\mathbf{q} \overset{\mathbf{E}}{\mapsto} \mathbf{q} - \mathbf{e}(\mathbf{p}, \mathbf{E}), \quad \mathbf{e}(\mathbf{p}, \mathbf{E}) = \sqrt{1 - p^2/n^2}\left(\mathbf{E} + \frac{\mathbf{p}\cdot\mathbf{E}}{1 - p^2/n^2}\,\mathbf{p}\right), \tag{6.6a}$$

$$\mathbf{p} \overset{\mathbf{E}}{\mapsto} \mathbf{p}, \qquad \text{and note } \mathbf{e}(\mathbf{p}, \mathbf{E}_1) + \mathbf{e}(\mathbf{p}, \mathbf{E}_2) = \mathbf{e}(\mathbf{p}, \mathbf{E}_1 + \mathbf{E}_2). \tag{6.6b}$$

For momentum translations (6.4*b*),

$$\begin{aligned} \mathbf{q} \overset{\mathbf{F}}{\mapsto} & \sqrt{1 + 2\mathbf{p}\cdot\mathbf{f}/n^2 + f^2/n^2} \\ & \times \left[\mathbf{q} + \tfrac{1}{n^2}\left\{\mathbf{f}\cdot(\mathbf{q} - \tfrac{1}{n^2}\mathbf{p}\cdot\mathbf{q}\,\mathbf{p})(\mathbf{p} + \mathbf{f}) + \mathbf{p}\cdot\mathbf{q}(1 - p^2/n^2)\mathbf{f}\right\}\right], \end{aligned} \tag{6.7a}$$

$$\mathbf{p} \overset{\mathbf{F}}{\mapsto} \frac{\mathbf{p} + \mathbf{f}}{\sqrt{1 + 2\mathbf{p}\cdot\mathbf{f}/n^2 + f^2/n^2}}, \quad \mathbf{f}(p^2, \mathbf{F}) = \sqrt{1 - p^2/n^2}\mathbf{F}. \tag{6.7b}$$

The translation (6.6) affects only the screen coordinate; for vanishingly small $\mathbf{p}$, $\mathbf{e}(\mathbf{p}, \mathbf{E}) \approx \mathbf{E}$ is the translation (6.6*a*) of $\mathbf{q}$; as $|\mathbf{p}|$ becomes larger, $\mathbf{E}$ starts to differ from $\mathbf{e}(\mathbf{p}, \mathbf{E})$; as $|\mathbf{p}| \to n$, the translation diverges in the direction of $\mathbf{p}$. The momentum translation (6.7) is more complex since it affects both $\mathbf{q}$ and $\mathbf{p}$. Again, for vanishingly small $|\mathbf{p}|$, $\mathbf{f}(p^2, \mathbf{F}) \approx \mathbf{F}$; the size of the vector $\mathbf{f}$ decreases with increasing $p^2 < n^2$; as $|\mathbf{p}|$ approaches n, the Euclidean limit of large-angle rays, we have $\mathbf{f}(\mathbf{p}, \mathbf{F}) = \mathbf{0}$. This effect may be seen in Figure 2 as we translate the Heisenberg-Weyl momentum plane $\mathbf{P}$: the $\mathbf{p}$-sphere moves as the image in Escher's reflecting balls. It is important to note that both (6.6) and (6.7) have the functional dependence of $\mathbf{p}$ and $\mathbf{q}$ given by a *point* transformation in $\mathbf{p}$, *i.e.*, $\mathbf{p}'(\mathbf{p})$ but $\mathbf{q}'(\mathbf{p}, \mathbf{q})$. This is a *distortion* in momentum space that entails a *comatic* aberration in the image on the screen.

We can now write the linear transformation $\mathcal{S}$ in (6.5), generated by the quadratic functions (6.2) on the true Euclidean optical phase space variables $\mathbf{p}$, $\mathbf{q}$. For the rotation subalgebra and subgroup, the results will be given in the next Section. Here we can write the results for the $\mathrm{Sp}(2, \Re)$ action of Heisenberg-Weyl axis-symmetric optical systems generated by (6.3).

Thus:

$$\begin{aligned}\begin{pmatrix}\mathbf{p}\\ \mathbf{q}\end{pmatrix} &\mapsto \mathcal{C}^{-1}\mathcal{S}\mathcal{C}\begin{pmatrix}\mathbf{p}\\ \mathbf{q}\end{pmatrix} \qquad (6.8a)\\ &= \mathcal{C}^{-1}\mathcal{S}\begin{pmatrix}\mathbf{p}/\sqrt{1-p^2/n^2}\\ \sqrt{1-p^2/n^2}(\mathbf{q}-\mathbf{p}\cdot\mathbf{q}\,\mathbf{p})\end{pmatrix}\\ &= \mathcal{C}^{-1}\begin{pmatrix}(A\mathbf{p}+B\mathbf{q})/R_{\mathbf{S}}(\mathbf{p},\mathbf{q})\\ R_{\mathbf{S}}(\mathbf{p},\mathbf{q})[(C\mathbf{p}+D\mathbf{q})-T_{\mathbf{S}}(\mathbf{p},\mathbf{q})(A\mathbf{p}+B\mathbf{q})]\end{pmatrix}\\ &= \begin{pmatrix}(A\mathbf{p}^c+B\mathbf{q}^c)/R^c\\ R^c([C-AT^c_{\mathbf{S}}]\mathbf{p}^c+[D-BT^c_{\mathbf{S}}]\mathbf{q}^c)\end{pmatrix} \qquad (6.8b)\end{aligned}$$

where $\mathbf{p}^c = \mathcal{C}^{-1}\mathbf{p}$ and $\mathbf{q}^c = \mathcal{C}^{-1}\mathbf{q}$ are given by (4.6), while $R^c(\mathbf{p},\mathbf{q}) = R(\mathbf{p}^c,\mathbf{q}^c)$ and $T^c_{\mathbf{S}}(\mathbf{p},\mathbf{q}) = T_{\mathbf{S}}(\mathbf{p}^c,\mathbf{q}^c)$ are to be replaced from

$$\begin{aligned}R_{\mathbf{S}}(\mathbf{p},\mathbf{q}) &= \sqrt{1-(A^2p^2+2AB\mathbf{p}\cdot\mathbf{q}+B^2q^2)/n^2}, \qquad (6.9a)\\ T_{\mathbf{S}}(\mathbf{p},\mathbf{q}) &= (2ACp^2+[AD+BC]\mathbf{p}\cdot\mathbf{q}+2BDq^2)/n^2. \qquad (6.9b)\end{aligned}$$

This is a case of a point-linear-point transformation (in N dimensions) of the kind studied by Leyvraz and Seligman [20], exchanging $\mathbf{p}$ and $\mathbf{q}$.

We can explicitly find the linear transformation subgroup generated by Heisenberg-Weyl free flight,

$$\mathcal{A}_\alpha\begin{pmatrix}\mathbf{P}\\ \mathbf{Q}\end{pmatrix} = \exp\alpha\widehat{P^2}\begin{pmatrix}\mathbf{P}\\ \mathbf{Q}\end{pmatrix} = \begin{pmatrix}\mathbf{P}\\ \mathbf{Q}-2\alpha\mathbf{P}\end{pmatrix} = \begin{pmatrix}1 & 0\\ -2\alpha & 1\end{pmatrix}\begin{pmatrix}\mathbf{P}\\ \mathbf{Q}\end{pmatrix}, \qquad (6.10a)$$

on the Euclidean phase space coordinates. It is

$$\begin{aligned}\mathcal{A}_\alpha\,\mathbf{p} &= \mathbf{p}, \qquad (6.10a)\\ \mathcal{A}_\alpha\,\mathbf{q} &= \mathbf{q}-2\alpha\frac{\mathbf{p}}{(1-p^2/n^2)^2}. \qquad (6.10b)\end{aligned}$$

For small $|\mathbf{p}|$, the relation of $\mathcal{A}_\alpha\,\mathbf{q}$ with $\mathbf{q}$ and $\mathbf{p}$ appears linear. For $|\mathbf{p}|$ approaching n, the screen transformation (6.10c) of $\mathbf{q}$ diverges in the direction of $-\mathbf{p}$ when $\alpha > 0$. This is the inverse of *spherical aberration, i.e.*, the error we incurr when we propagate Euclidean rays by Heisenberg-Weyl free flight.

Pure magnification,

$$\mathcal{B}_\beta\begin{pmatrix}\mathbf{P}\\ \mathbf{Q}\end{pmatrix} = \exp\beta\widehat{\mathbf{P}\cdot\mathbf{Q}}\begin{pmatrix}\mathbf{P}\\ \mathbf{Q}\end{pmatrix} = \begin{pmatrix}e^\beta\mathbf{P}\\ e^{-\beta}\mathbf{Q}\end{pmatrix} = \begin{pmatrix}e^\beta & 0\\ 0 & e^{-\beta}\end{pmatrix}\begin{pmatrix}\mathbf{P}\\ \mathbf{Q}\end{pmatrix}, \qquad (6.11a)$$

on Euclidean phase space is

$$\begin{aligned}\mathcal{B}_\beta\,\mathbf{p} &= \frac{e^\beta\mathbf{p}}{\sqrt{1+(e^{2\beta}-1)p^2/n^2}}, \qquad (6.11a)\\ \mathcal{B}_\beta\,\mathbf{q} &= \sqrt{1+(e^{2\beta}-1)p^2/n^2}\Big[e^{-\beta}\mathbf{q}+2\frac{1}{n^2}\sinh\beta\,\mathbf{p}\cdot\mathbf{q}\,\mathbf{p}\Big]. \qquad (6.11b)\end{aligned}$$

Again we check that for $|\mathbf{p}| \ll n$ the map resembles a linear one, while for $|\mathbf{p}| \to n$ the action becomes the identity, keeping the boundary of the $|\mathbf{p}| = n$ sphere in its place. Again, a glance at Figure 2 confirms intuition.

Finally, a "Gaussian thin lens" transformation,

$$\mathcal{C}_\gamma \begin{pmatrix} \mathbf{P} \\ \mathbf{Q} \end{pmatrix} = \exp \gamma \widehat{Q^2} \begin{pmatrix} \mathbf{P} \\ \mathbf{Q} \end{pmatrix} = \begin{pmatrix} \mathbf{P} + 2\gamma \mathbf{Q} \\ \mathbf{Q} \end{pmatrix} = \begin{pmatrix} 1 & 2\gamma \\ 0 & 1 \end{pmatrix} \begin{pmatrix} \mathbf{P} \\ \mathbf{Q} \end{pmatrix}, \qquad (6.12a)$$

on the full direction hemisphere appears as

$$\mathcal{C}_\gamma \, \mathbf{p} = \frac{\mathbf{p}_\gamma}{\sqrt{1 - p^2/n^2 + p_\gamma^2/n^2}}, \qquad (6.12b)$$

where

$$\mathbf{p}_\gamma(\mathbf{p}, \mathbf{q}, \gamma) = \mathbf{p} + 2\gamma(1 - p^2/n^2)(\mathbf{q} - \frac{1}{n^2}\mathbf{p}\cdot\mathbf{q}\,\mathbf{p}), \qquad (6.12c)$$

and

$$\mathcal{C}_\gamma \, \mathbf{q} = \sqrt{1 - p^2/n^2 + p_\gamma^2/n^2}\left(1 + \frac{\mathbf{p}_\gamma \mathbf{p}_\gamma \cdot}{n^2 - p^2}\right)\left(\mathbf{q} - \frac{1}{n^2}\mathbf{p}\cdot\mathbf{q}\,\mathbf{p}\right). \qquad (6.12d)$$

The function $\mathbf{p}_\gamma(\mathbf{p}, \mathbf{q}, \gamma)$ in (6.12*c*) becomes linear for small $|\mathbf{p}|$ and approaches the identity at n. The phase space transformations in (6.12*b*) and (6.12*d*) have the same behaviour; the boundary $|\mathbf{p}| = n$ stays put.

Much, if not the whole Lie theory of aberrations, has thus far been based on transformations generated by polynomial functions of the Heisenberg-Weyl phase space variables [1], [6]. We may now let these act on the phase space of Euclidean optics. Thus far we have involved only the *forward* hemisphere of ray directions, $h > 0$. If we take the negative sign of the square roots in (4.3) we open the Euclidean momentum sphere onto a *second* Heisenberg-Weyl momentum plane that describes *backward*-moving rays.

7.7 The Euclidean and Lorentz groups on Heisenberg-Weyl phase space

The natural group of motions of Euclidean optical phase space is precisely the Euclidean group. For the general case of $(N + 1)$-dimensional optics with N-dimensonal screens this is the group ISO$(N + 1)$. We shall now discuss how this group acts on the $\Re^{2N}$ phase space of Heisenberg-Weyl optics, why *two* copies of the momentum space $\Re^N$ space are required, and a certain similarity group that seems to be operating.

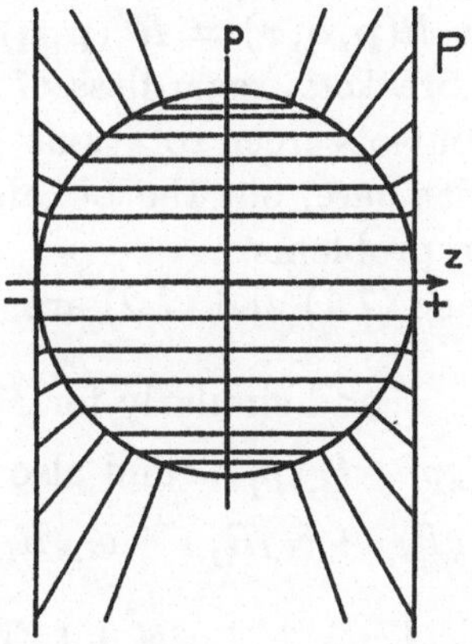

FIGURE 5. The map of the sphere of ray directions between two Euclidean disks and two Heisenberg-Weyl planes.

The N functions that generate translations in the Euclidean screen and the $(N+1)^{\text{th}}$ generator of evolution *normal* to the screen are

$$p_i = \frac{P_i}{\sqrt{1+P^2/n^2}}, \qquad i = 1,2,\ldots,N, \tag{7.1a}$$

$$h^\tau = \tau\sqrt{n^2-p^2} = \frac{\tau n}{\sqrt{1+P^2/n^2}}, \qquad \tau = -1,0,+1. \tag{7.1b}$$

The last generator is (minus) the Hamiltonian; its two signs give the two forms of the generator on each of the two Euclidean momentum spheres (*disks* in ordinary $N=2$ optics), corresponding to forward- and backward-directed rays.[7] The latter are the mirror images of the first reflected on the reference plane, exhibiting the same value of the momentum $\mathbf{p}$, but differing in behaviour under z-translations due to the sign τ of h, the label of the *chart*. See Figure 5.

The generators of rotations *within* the Euclidean screen submanifold are

$$R_{i,j} = p_i q_j - p_j q_i = P_i Q_j - P_j Q_i, \qquad i \neq j = 1,2,\ldots N. \tag{7.2a}$$

The forms are the same in both spaces because they depend only on the *angular* transformation properties of $\{\mathbf{p},\mathbf{q}\}$ and $\{\mathbf{P},\mathbf{Q}\}$, and not of their sizes, nor τ; they close into the Lie algebra of SO(N).

Rotations *out* of the plane of the screen are generated by

$$R^\tau_{i,N+1} = q_i h^\tau = \tau(nQ_i + \tfrac{1}{n}\mathbf{P}\cdot\mathbf{Q}\,P_i), \qquad i = 1,2,\ldots N. \tag{7.2b}$$

[7] Strictly, we should speak of forward and backward hemispheres, joined by an *equatorial circle*, corresponding to $\tau = +1$, -1 and 0. The equatorial rays are disregarded here; they should serve as limit points for sequences of rays in either hemisphere [10].

These are *two-chart* functions $R(\mathbf{p},\mathbf{q};\tau) = R^\tau(\mathbf{p},\mathbf{q}) = \tau R^{+1}(\mathbf{p},\mathbf{q})$. They formally close under Poisson brackets, regardless of the fact that the two charts are mixed under finite out-of-screen rotations: the corresponding Lie exponentials must be dealt with care, but the Lie algebraic properties are less sensitive to global domain problems.

The Poisson brackets between (7.1) and (7.2) are:

$$\begin{aligned} \{p_i, p_j\} &= 0, \qquad \text{and similarly for } P\text{'s}, && (7.3a)\\ \{R_{i,j}, p_k\} &= \delta_{j,k}p_i - \delta_{i,k}p_j, \quad \text{and also } P\text{'s}, && (7.3b)\\ \{R_{i,j}, R_{k,l}\} &= \delta_{j,k}R_{i,l} + \delta_{i,l}R_{j,k} + \delta_{i,k}R_{l,j} + \delta_{j,l}R_{k,i}. && (7.3c) \end{aligned}$$

The Lie brackets hold for $i, \ldots, l = 1, 2, \ldots, N+1$. The first two also hold when we replace p_i by q_i for $i = 1, 2, \ldots, N$. The functions q_i, however, are not among the *generators* of the Euclidean group of motions; the handling of q_i's by themselves follows from the canonical formalism. In Ref. [10] we presented arguments to distrust the Euclidean wavization of the q_i into a self-adjoint operator. In the Heisenberg-Weyl setting, on the other hand, the natural and consistent quantization of Q_i is into the familiar self-adjoint Schrödinger operator.

The N-sphere of ray directions of Euclidean optics lends itself for the action of the Lorentz group SO($N+1$,1). This group of relativity is a *deformation* of the basic Euclidean group ISO($N+1$) [12]. The boost generators in N dimensions are given by the $(N+1)$-vector

$$\begin{aligned} B_i &= nq_i - \tfrac{1}{n}\mathbf{p}\cdot\mathbf{q}\,p_i = \sqrt{n^2+P^2}\,Q_i, && (7.4a)\\ B^\tau_{N+1} &= \quad -\tfrac{1}{n}\mathbf{p}\cdot\mathbf{q}\,h^\tau = -\tau\sqrt{1+P^2/n^2}\,\mathbf{P}\cdot\mathbf{Q}. && (7.4b) \end{aligned}$$

The brackets between (7.4) and (7.2) close into the Lorentz algebra:

$$\begin{aligned} \{R_{i,j}, B_k\} &= \delta_{j,k}B_i - \delta_{i,k}B_j, && (7.5a)\\ \{B_i, B_j\} &= -R_{i,j}. && (7.5b) \end{aligned}$$

The last function (7.4*b*) generates boosts along the optical axis; it is also a two-chart function due to the factor h^τ containing τ; we shall omit the index τ on h since it is given by the latter's sign. The Lie transformation $\exp\alpha\hat{B}_{N+1}$ distorts the angle θ (between a ray and the optical axis) through [12] $\tan\frac{1}{2}\theta \mapsto \tan\frac{1}{2}\theta' = e^{-\alpha}\tan\frac{1}{2}\theta$, mapping thus the backward direction region $\frac{1}{2}\pi \le \theta < 2\arctan e^{\alpha}$ into the forward hemisphere.

We should point out the striking similarity between the expressions for out-of-screen rotations (7.2*b*) in terms of $\mathbf{p}$ and $\mathbf{q}$, and that of boosts (7.4*a*) in $\mathbf{P}$ and $\mathbf{Q}$. Indeed, $R_{i,N+1}(q_i, \sqrt{1-p^2/n^2}) = B_i(Q_i, \sqrt{1+P^2/n^2})$. The R's close under Poisson bracket to a rotation algebra and the B's to a Lorentz algebra because of the sign of p^2 in the square root is negative while that of P^2 is positive.

The Euclidean translations are generated by (7.1). The Lie transformation $\stackrel{\mathbf{e}}{\mapsto}$ of translation by $\mathbf{e} = (e_1, e_2, \ldots, e_N)$ in the screen plane is $\exp \sum e_i \hat{p}_i$. Its action on Heisenberg-Weyl phase space is the converse of (6.6), namely

$$\mathbf{Q} \stackrel{\mathbf{e}}{\mapsto} \mathbf{Q} - \mathbf{E}(\mathbf{P}, \mathbf{e}), \quad \mathbf{E}(\mathbf{P}, \mathbf{e}) = \frac{1}{\sqrt{1 + P^2/n^2}} \left(\mathbf{e} + \frac{\mathbf{P} \cdot \mathbf{e}}{n^2 + P^2} \mathbf{P} \right) \quad (7.6a)$$

$$\mathbf{P} \stackrel{\mathbf{e}}{\mapsto} \mathbf{P}, \qquad \text{and note } \mathbf{E}(\mathbf{P}, \mathbf{e}_1) + \mathbf{E}(\mathbf{P}, \mathbf{e}_2) = \mathbf{E}(\mathbf{P}, \mathbf{e}_1 + \mathbf{e}_2). \quad (7.6b)$$

Euclidean rotations *in* the N-screen, generated by $R_{i,j}$ in (7.2a), commute with the opening coma map, and provide the Petzval invariant as $\frac{1}{2} \sum_{i,j} R_{i,j} R_{i,j}$, the SO($N$) second-order Casimir operator. These rotations coincide with the maximal compact subgroup SO(N) $\subset$ Sp($2N, \Re$). On the other hand, SO($N + 1$) rotations (7.2b) *out* of the screen plane, are generated by $\sum \alpha_i R^T_{i,N+1}$. For any single i, the effect of the Lie transformation $\mathcal{R}_i(\alpha) = \exp \alpha \hat{R}^T_{i,N+1} = \exp \alpha \widehat{q_i h}$ on Euclidean ray direction is

$$\mathcal{R}_i(\alpha) : p_j \;=\; \begin{cases} p_i \cos\alpha + h \sin\alpha, & i = j, \\ p_j, & i \neq j, \end{cases} \quad (7.7a)$$

$$\mathcal{R}_i(\alpha) : h \;=\; -p_i \sin\alpha + h \cos\alpha. \quad (7.7b)$$

In writing h explicitly in the transformation we can distinguish between the forward and backward hemispheres through its sign.

To obtain the action on Euclidean ray position q_i, we note that the transformation properties (7.7) of the p_j are the same as those of $R^T_{j,N+1} = q_i h$, hence

$$\mathcal{R}_i(\alpha) : R_{i,N+1} = \begin{cases} R_{i,N+1}, & i = j \\ R_{i,N+1} \cos\alpha + R_{i,j} \sin\alpha, & i \neq j. \end{cases} \quad (7.8)$$

Thus, writing $R_{i,N+1} = q_i h \mapsto R'_{i,N+1} = q'_i h'$, we find the effect of the Lie transformation on Euclidean position as

$$\mathcal{R}_i(\alpha) : q_j = \begin{cases} \dfrac{q_i}{\cos\alpha - p/h \,\sin\alpha,} & i = j, \\ \dfrac{q_j \cos\alpha - (p_i q_j - p_j q_i)/h \,\sin\alpha}{\cos\alpha - p_i/h \,\sin\alpha} & i \neq j. \end{cases} \quad (7.9)$$

The denominator may vanish for $\tan\alpha = h/p_i$: rays that become paralell to the screen. The singularity only indicates that a change of chart is required to go beyond, because then the ray will come from the opposite side of the screen.

Let us write explicitly the Euclidean action on the Heisenberg-Weyl variables $\{\mathbf{P}, \mathbf{Q}\} \in \Re^{2N}$, obtained through the opening coma map. Translations generated by $p_i(\mathbf{P})$ in (7.1) were shown to have the action (7.6). The SO(N)

rotations generated by (7.2a) have the same action on $\{\mathbf{P}, \mathbf{Q}\} \in \Re^{2N}$ as on $\{\mathbf{p}, \mathbf{q}\} \in \Re^{N}_{p<n} \otimes \Re^{N}$. Out-of-screen rotations in are SO($N+1$) generated by (7.2b) and their action on p-space was seen in (7.7a) and on q-space in (7.9). Aiding ourselves with the transformation (7.7b) of h as component $N+1$ in Euclidean optics, we find for the Heisenberg-Weyl momentum space

$$\mathcal{R}_i(\alpha) : P_j = \begin{cases} \dfrac{P_i \cos\alpha + n \sin\alpha}{\cos\alpha - P_i/n \sin\alpha}, & i = j, \\ \dfrac{P_j}{\cos\alpha - P_i/n \sin\alpha} & i \neq j, \end{cases} \tag{7.10a}$$

$$\mathcal{R}_i(\alpha) : \sqrt{1 + P^2/n^2} = \frac{\sqrt{1+P^2/n^2}}{\cos\alpha - P_i/n \sin\alpha}. \tag{7.10b}$$

To obtain the action on position Q_i through the opening coma map (4.3b), the rotation on q_j from (7.9) and the inverse coma map (4.6b), it is useful to know the transformation properties of $\mathbf{p}\cdot\mathbf{q}$ under such rotations. From (7.7) and (7.9), and in view of (4.8) and (7.10b),

$$\mathcal{R}_i(\alpha) : \mathbf{p}\cdot\mathbf{q} = \frac{\mathbf{p}\cdot\mathbf{q} \cos\alpha + n^2/h\, (q_i - \frac{1}{n^2}\mathbf{p}\cdot\mathbf{q}\, p_i) \sin\alpha}{\cos\alpha - p_i/h \sin\alpha}. \tag{7.11}$$

The expression in parentheses above is $q_i - \frac{1}{n^2}\mathbf{p}\cdot\mathbf{q}\, p_i = Q_i\sqrt{1+P^2/n^2}$. Through this we obtain

$$\mathcal{R}_i(\alpha) : Q_j = \begin{cases} (\cos\alpha - P_i/n \sin\alpha)(Q_i \cos\alpha - \frac{1}{n}\mathbf{P}\cdot\mathbf{Q} \sin\alpha), & i = j, \\ Q_j (\cos\alpha - P_i/n \sin\alpha), & i \neq j, \end{cases} \tag{7.12a}$$

$$\mathcal{R}_i(\alpha) : \mathbf{P}\cdot\mathbf{Q} = (\cos\alpha - P_i/n \sin\alpha)(-nQ_i \sin\alpha + \mathbf{P}\cdot\mathbf{Q} \cos\alpha). \tag{7.12b}$$

We write the last expression because the out-of-plane rotations $\mathcal{R}_i(\alpha)$ will transform the components of B_i, $i = 1, 2, \ldots, N, N+1$ amongst themselves;[8] modulo the factor $\sqrt{n^2+P^2}$, $\mathbf{P}\cdot\mathbf{Q}/n$ suggests itself as the $(N+1)^{\text{th}}$ component of the N-vector $\mathbf{Q}$. The $(N+1)^{\text{th}}$ component of the Euclidean N-vector $\mathbf{q}$ is zero, as can be seen in the same equations (7.3) because we are in the standard Euclidean screen. In a similar way, the factor $H = \sqrt{n^2+P^2}$ transforms as the $(N+1)^{\text{th}}$ component of the N-vector $\mathbf{P}$, as was given in Eq. (7.10b); this fittingly corresponds with h, *cf.* Eqs. (4.7), as the $(N+1)^{\text{th}}$ component of the $(N+1)$-vector of ray direction on the Descartes sphere $p^2 + h^2 = n^2$. We are now on a *hyperboloid* $H^2 - P^2 = n^2$, however.

---

[8]Indeed, the B_i may be written as $R_{i,N+2}$ and appended to the list (7.2) with commutation relations (7.3c), except that the Kronecker δ's that appear should be replaced by $g_{i,j}$ that is 1 when $i = j \leq N$ and -1 when $i = j \geq N+1$, and zero otherwise.

To obtain the action of the Lorentz boost operators (7.4) on Euclidean and Heisenberg-Weyl rays, we use the curious similarity noted above. Indeed, transformations generated by $B_i = nq_i - \frac{1}{n}\mathbf{p}\cdot\mathbf{q}\,p_i$ on $\{\mathbf{p},\mathbf{q}\}$ are 'similar' to those generated by $R_{i,N+1} = nQ_i + \frac{1}{n}\mathbf{P}\cdot\mathbf{Q}\,P_i$ on $\{\mathbf{P},\mathbf{Q}\}$, the latter given by (7.10) and (7.12). Also, transformations generated by $B_i = HQ_i$ on $\{\mathbf{P},\mathbf{Q}\}$ will be similar to $R_{i,N+1} = hq_i$ acting on $\{\mathbf{p},\mathbf{q}\}$, in (7.7) and (7.9). Replacing $\{q_i, h\} = -p_i/h$ by $\{Q_i, H\} = +P_i/H$ seems to have the effect of changing the series $\sin\alpha \mapsto -\sinh\alpha$ and $\cos\alpha \mapsto \cosh\alpha$. As a check, we may compare $\mathcal{R}_i(\alpha) : Q_j$ in (7.12) through the upper-by-lower-case and series replacements with equations (13.6) in reference [10], found in [12].

We are left with the computation of the transformation generated by the boost normal to the screen, $B_{N+1} = -\frac{1}{n}\mathbf{p}\cdot\mathbf{q}\,h = -\mathbf{P}\cdot\mathbf{Q}\,H$. This is self-similar in the same sense as the $R_{i,j}$ in (7.2a). The boost formulas appear in the companion article [10], Eqs. (13.1) and (13.3), and were also presented in [12]. For completeness, we reproduce the formulas here for the Euclidean phase space variables:

$$\mathcal{B}_z : \mathbf{p} = \frac{\mathbf{p}}{\cosh\alpha + h/n\,\sinh\alpha}, \qquad (7.13a)$$

$$\mathcal{B}_z : \mathbf{q} = (\cosh\alpha + h/n\,\sinh\alpha)\left(\mathbf{q} - \frac{\sinh\alpha}{n\sinh\alpha + h\cosh\alpha}\,\frac{\mathbf{p}\cdot\mathbf{q}}{n}\,\mathbf{p}\right) \qquad (7.13b)$$

The same transformation holds for the Heisenberg-Weyl variables upon replacement of $\mathbf{q}$, $\mathbf{p}$,and h by $\mathbf{Q}$, $\mathbf{P}$, and $H = \sqrt{n^2 + P^2}$.

7.8 Spherical aberration, coma, and point transformations in phase space

Let us remark a general feature of the maps seen here and, indeed, of the generator functions themselves. They have the structure $P \mapsto f(P)$, $Q \mapsto g(P)Q + h(P)$ and similarly for the lower-case Euclidean variables, and in many dimensions. Such are *point* transformations in P-space, a proper subset of the group of *all* canonical transformations in phase space. The generators of $(2m-1)^{\text{th}}$-order spherical aberration [6] are $(P^2)^m$, $m = 1, 2, \ldots$, while those of $(2m-1)^{\text{th}}$-order circular coma[9] are $(P^2)^{m-1}\mathbf{P}\cdot\mathbf{Q}$.

The Lie transformations generated by (7.1b), or by any functions $f(p^2) = F(P^2)$, will map $\{\mathbf{P},\mathbf{Q}\} \in \Re^{2N}$ in the following way:

$$\exp A\hat{F}(P^2)\begin{pmatrix}\mathbf{P}\\ \mathbf{Q}\end{pmatrix} = \begin{pmatrix}\mathbf{P}\\ \mathbf{Q} - 2A\,F'(P^2)\mathbf{P}\end{pmatrix}. \qquad (8.1a)$$

---

[9]For $m = 1$ we have first-order 'aberrations'. The linear symplectic 'free flight' P^2 is first-order spherical aberration; magnification $\mathbf{P}\cdot\mathbf{Q}$ is first-order coma and, as its own Fourier transform, first-order distortion.

In Optics this is termed *spherical aberration* because a pencil of rays $\mathbf{P} \in \Re^2$ issuing from an object at $\mathbf{Q}_0$, produces an image $\mathbf{Q}(\mathbf{P}, \mathbf{Q}_0) = \mathbf{Q}_0 + \alpha\mathbf{\Phi}(\mathbf{P})$ containing a spread $\mathbf{\Phi}$ paralell to $\mathbf{P}$, unbounded, and independent of the location of the source $\mathbf{Q}_0$ in the object screen.

The generators (7.1a) of translations *in* the screen are of the form $F(P^2)\mathbf{P}$. The Lie transformation generated by a coefficient vector $\mathbf{B}$ is

$$\exp[F(P^2)\,\mathbf{B}\cdot\mathbf{P}]^{\wedge}\begin{pmatrix}\mathbf{P}\\ \mathbf{Q}\end{pmatrix} = \begin{pmatrix}\mathbf{P}\\ \mathbf{Q}-[F+2F'\,\mathbf{P}\mathbf{P}\cdot]\mathbf{B}\end{pmatrix}. \tag{8.1b}$$

Rotations generated by (7.2a) will transform transformations (8.1b) among themselves.

In-screen rotations (7.2a) are standard, but (7.2b) and (7.4) are of the generic form $g(\mathbf{p},\mathbf{q}) = g_i^I(p^2)q_i + g_i^{II}(p^2)\mathbf{p}\cdot\mathbf{q}\,p_i = G(\mathbf{P},\mathbf{Q}) = G_i^I(P^2)Q_i + G_i^{II}(P^2)\mathbf{P}\cdot\mathbf{Q}\,P_i$, while the SO($N$) scalar (7.4$b$) has the form $h(p^2)\mathbf{p}\cdot\mathbf{q} = H(P^2)\mathbf{P}\cdot\mathbf{Q}$. They are *linear* in the components of position: $g(\mathbf{p}, x\mathbf{q}) = xg(\mathbf{p},\mathbf{q})$ and $G(\mathbf{P}, X\mathbf{Q}) = XG(\mathbf{P},\mathbf{Q})$. The Lie transformations they generate are of the form

$$\exp \mathbf{C}\cdot\hat{\mathbf{G}}\begin{pmatrix}\mathbf{P}\\ \mathbf{Q}\end{pmatrix} = \begin{pmatrix}\mathbf{P}'(\mathbf{P})\\ \mathbf{Q}'(\mathbf{P},\mathbf{Q};\mathbf{C})\end{pmatrix}, \tag{8.2}$$

where $\mathbf{Q}'(\mathbf{P},\mathbf{Q};\mathbf{C})$ is a function of the *same* kind as G described above with vector indices 'balanced' to one. Similarly for the lower-cased letters. These are *point* transformations of momentum space (*i.e.*, *distortions* of it). As far as (7.4b) is concerned, and as detailed in Refs. [10] and [12], the Fourier transform of distortion is circular comatic aberration of the image position space. The spot diagrams of Lorentz boosts (7.4a) in the screen plane (Ref.[10], Fig. 5b), are SO($N+1$)–*rotated* versions of the basic comatic aberration (in Ref.[10], Figs. 5a and 5c).

For $N = 1$ dimension, it is well known that functions $\Phi(P,Q) = \Phi^I(P)Q + \Phi^{II}(P)$ close under Poisson brackets and thus generate an infinite-dimensional subalgebra in the enveloping algebra of Heisenberg and Weyl. They thus serve also as function space for the Lie transformations. In addition, such functions are uniquely quantized [7]. [10] For $N = 1$ dimension, we have explicit formulas (with an indefinite integral) for $Q'(P,Q;C)$; singularities may occur, having to do with the Φ's range. General formulas for the N-dimensional case are not available, but discussions with F. Leyvraz [20], IF–UNAM, suggest that they can be found explicitly.

[10] For such functions, Poisson brackets and commutators of their Schrödinger operators follow each other. All quantization rules leading to self-adjoint operators give the same result on functions that are linear in one of the conjugate observables.

Spherical aberration and circular coma (of all orders), together with their rotated aberrations and the opening coma introduced here, constitute a subring of Heisenberg-Weyl whose study should be interesting. The Euclidean and Lorentz algebra presented above are only (the only?) finite-dimensional subalgebras.

7.9 The Hilbert spaces for Heisenberg-Weyl and Euclidean optics

Quantization in the Heisenberg-Weyl model of optical phase space is essentially $\mathcal{L}_2(\Re^N)$ quantization à la Schrödinger. The Fourier transform plays a prominent role intertwining the configuration and momentum representations [8], so the subject is also called Fourier optics. Wavization of Euclidean optics, on the other hand, involves solutions of the wave equation; Fourier frequency analysis of a signal decomposes the space of waves into subsets that are solutions of the Helmholtz equation for each wavenumber [10]. There is a 'Helmholtz-Hilbert space' that uses for function range and inner product integration the two momentum spheres (*disks* in $N = 2$ dimensions [10]), joined at their surfaces (perimeter in the $N = 2$ case). Here we shall show how the opening coma transformation intertwines between functions in Heisenberg-Weyl and Euclidean-Helmholtz wave optics.

The opening coma map (4.3*a*)–(4.6*a*) in radial and angular variables, is

$$P = \frac{p}{\sqrt{1 - p^2/n^2}}, \qquad \Omega_{\mathbf{P}} = \omega_{\mathbf{p}} \in \mathcal{S}_{N-1}, \tag{9.1a}$$

and we also need the radial differential

$$dP = (1 - p^2/n^2)^{-3/2} dp. \tag{9.1b}$$

We express the inner product of the $\mathcal{L}_2(\Re^N)$ Hilbert space of square-integrable Heisenberg-Weyl wavefunctions in the momentum representation $\mathbf{P} \in \Re^N$, in terms of an integral of the Euclidean momentum $\mathbf{p}$ over *its* proper range $|\mathbf{p}| < n$. Thus,

$$(\Phi, \Psi)_{\mathcal{L}_2(\Re^N)} = \int_{\Re^N} d^N\mathbf{P}\, \Phi(\mathbf{P})^* \Psi(\mathbf{P}) \tag{9.2a}$$

$$= \int_0^\infty P^{N-1} dP \times \int_{\mathcal{S}_{N-1}} d^{N-1}\Omega_{\mathbf{P}}\, \Phi(P, \Omega_{\mathbf{P}})^* \Psi(P, \Omega_{\mathbf{P}}) \tag{9.2b}$$

$$= \int_0^n \frac{p^{N-1}\, dp}{(1 - p^2/n^2)^{N/2+1}} \times \int_{\mathcal{S}_{N-1}} d^{N-1}\omega_{\mathbf{p}}\, \Phi(\mathbf{P}(p, \omega_{\mathbf{p}}))^* \Psi(\mathbf{P}(p, \omega_{\mathbf{p}})) \tag{9.2c}$$

$$= \int_{\substack{\mathbf{p}\in\Re^N \\ |\mathbf{p}|<n}} \frac{d^N\mathbf{p}}{(1-p^2/n^2)^{N/2+1}} \Phi(\mathbf{P}(\mathbf{p}))^* \Psi(\mathbf{P}(\mathbf{p})) \quad (9.2d)$$

$$= \int_{\substack{\mathbf{p}\in\Re^N \\ |\mathbf{p}|<n}} \frac{d^N\mathbf{p}}{\sqrt{1-p^2/n^2}} \phi(\mathbf{p})^* \psi(\mathbf{p}) \quad (9.2e)$$

$$= \int_{\mathcal{S}_N^{(+)}} d^N S(\vec{p})\, \phi(\vec{p})^*\, \psi(\vec{p}) = (\phi, \psi)_{\mathcal{L}_2(\mathcal{S}_N^{(+)})}. \quad (9.2f)$$

The $\mathcal{L}_2(\Re^N)$ inner product (9.2a) is expressed in the radial and angular variables of $\mathbf{P}$ in (9.2b) and of $\mathbf{p}$ in (9.2c). The integral (9.2d) shows that the inner product in Euclidean momentum variables has a weight factor $(1-p^2/n^2)^{-(N/2+1)}$. The integrand in (9.2e) defines ϕ and ψ such that the last integral, (9.2f), be over the *forward hemisphere* $\mathcal{S}_N^{(+)}$ of the N-dimensional surface of the *Descartes* sphere of ray directions. The sphere $\mathcal{S}_N$ is inmersed in an $(N+1)$-dimensional space whose *ray* vectors are $\vec{p} = \{\mathbf{p} = n\sin\theta\,\omega_\mathbf{p},\ h = n\cos\theta > 0\}$, whose radii are $\vec{p}\cdot\vec{p} = p^2 + h^2 = n^2$, and we may write $p = |\mathbf{p}| = n\sin\theta$. The surface element of the Descartes sphere is then

$$d^N S(\vec{p}) = n^N \sin^{N-1}\theta\, d\theta\, d^{N-1}\omega_\mathbf{p} = \frac{p^{N-1}\,dp\, d^{N-1}\omega_\mathbf{p}}{\sqrt{1-p^2/n^2}} = \frac{d^N\mathbf{p}}{\sqrt{1-p^2/n^2}}. \quad (9.3)$$

In $\mathcal{S}_N^{(+)}$ the lower-case Greek wavefunctions of $\mathbf{p}$ are related to their upper-case counterparts through a measure normalization factor,

$$\begin{aligned} \psi(\vec{p}) &= \psi_+(\mathbf{p}) = |1-p^2/n^2|^{-(N+1)/4}\Psi(\mathbf{P}(\mathbf{p})) \\ &= |1+P^2/n^2|^{(N+1)/4}\Psi(\mathbf{P}(\mathbf{p})), \quad h = \sqrt{1-p^2/n^2} > 0. \quad (9.4a) \\ \Psi(\mathbf{P}) &= |1+P^2/n^2|^{-(N+1)/4}\psi_+(\mathbf{p}(\mathbf{P})) \\ &= |1-p^2/n^2|^{(N+1)/4}\psi_+(\mathbf{p}(\mathbf{P})). \quad (9.4b) \end{aligned}$$

In reference [10] we built the Hilbert space $\mathcal{L}_2(\mathcal{S}_N)$ on the *whole* Descartes sphere, natural subject to Euclidean transformations, and then projected it flat on *two* $|\mathbf{p}| < n$ disks sewn at the boundary. Equation (9.4a) defines the function $\psi(\vec{p})$ only on the forward $(h > 0)$ hemisphere. A second Heisenberg-Weyl $\Re^N$ space of fuctions $\Psi_b(\mathbf{P})$ is needed to map onto the *backward* $(h < 0)$ hemisphere by

$$\begin{aligned} \psi(\vec{p}) &= \psi_-(\mathbf{p}) = |1-p^2/n^2|^{(N+1)/4}\Psi_b(\mathbf{P}(\mathbf{p})) \\ &= |1+P^2/n^2|^{(N+1)/4}\Psi_b(\mathbf{P}(\mathbf{p})), \quad h = \sqrt{1-p^2/n^2} < 0. \quad (9.5a) \\ \Psi_b(\mathbf{P}) &= |1+P^2/n^2|^{-(N+1)/4}\psi_-(\mathbf{p}(\mathbf{P})) \\ &= |1-p^2/n^2|^{(N+1)/4}\psi_-(\mathbf{p}(\mathbf{P})). \quad (9.5b) \end{aligned}$$

7.10 Plane waves and the coma kernel

In Fourier-Heisenberg-Weyl optics, the image wave function on the screen is given by the *Fourier* transform of $\Psi(\mathbf{P})$,

$$\widetilde{\Psi}(\mathbf{Q}) = \left(\frac{k}{2\pi n}\right)^{N/2} \int_{\Re^N} d^N\mathbf{P}\,\Psi(\mathbf{P})\exp(ik\mathbf{P}\cdot\mathbf{Q}/n), \quad (10.1a)$$

$$\Psi(\mathbf{P}) = \left(\frac{k}{2\pi n}\right)^{N/2} \int_{\Re^N} d^N\mathbf{Q}\,\widetilde{\Psi}(\mathbf{Q})\exp(-ik\mathbf{P}\cdot\mathbf{Q}/n). \quad (10.1b)$$

The scale is given by the quantity k/n, of units of $[PQ]^{-1}$. The reduced wavelength $\lambda/2\pi = n/k$ is equivalent to $\hbar = h/2\pi$ in quantum mechanics. The inner product on the Fourier screen is, through the Parseval identity,

$$\begin{aligned}(\widetilde{\Phi},\widetilde{\Psi})_{\mathcal{L}_2(\Re^N)} &= \int_{\Re^N} d^N\mathbf{Q}\,\widetilde{\Phi}(\mathbf{Q})^*\,\widetilde{\Psi}(\mathbf{Q}) \\ &= \int_{\Re^N} d^N\mathbf{P}\,\Phi(\mathbf{P})^*\,\Psi(\mathbf{P}) = (\Phi,\Psi)_{\mathcal{L}_2(\Re^N)}. \quad (10.2)\end{aligned}$$

In Euclidean optics [10] we combine plane waves coming from all directions in the Descartes sphere. The basic linear combination function is $\psi(\vec{p})$ in (9.4a) and (9.5a), covering both backward and forward ray hemispheres. The wavefunction at the screen is then

$$\widetilde{\phi}(\vec{q})|_{q_N=0} = \left(\frac{k}{2\pi n}\right)^{N/2} \int_{S_N} d^N S(\vec{p})\,\phi(\vec{p})\exp(ik\vec{p}\cdot\vec{q}/n)\Big|_{q_N=0} \quad (10.3a)$$

$$= \widetilde{\phi}(\mathbf{q}) = \left(\frac{k}{2\pi n}\right)^{N/2} \int_{\substack{\mathbf{p}\in\Re^N\\|\mathbf{p}|<n}} \frac{d^N\mathbf{p}}{\sqrt{1-p^2/n^2}}[\phi_+(\mathbf{p})+\phi_-(\mathbf{p})] \times \exp(ik\mathbf{p}\cdot\mathbf{q}/n), \quad (10.3b)$$

and the *normal derivative* at the screen is

$$\frac{\partial\widetilde{\phi}(\vec{q})}{\partial q_N}\bigg|_{q_N=0} = \frac{ik}{n}\left(\frac{k}{2\pi n}\right)^{N/2} \times \int_{S_N} d^N S(\vec{p})\,p_N\,\phi(\vec{p})\exp(ik\vec{p}\cdot\vec{q}/n)\Big|_{q_N=0} \quad (10.3c)$$

$$= \widetilde{\phi}'(\mathbf{q}) = ik\left(\frac{k}{2\pi n}\right)^{N/2} \times \int_{\substack{\mathbf{p}\in\Re^N\\|\mathbf{p}|<n}} d^N\mathbf{p}\,[\phi_+(\mathbf{p})-\phi_-(\mathbf{p})]\exp(ik\mathbf{p}\cdot\mathbf{q}/n). \quad (10.3d)$$

The wavefunction $\widetilde{\phi}(\vec{q})$ is a solution of the Helmholtz equation $\sum_{i=1}^{N+1} \partial^2/\partial q_i^2 \; \phi(\vec{q}) = -k^2\phi(\vec{q})$ [10]. The inverse spectrum analysis of Helmholtz solutions (10.3*b*) requires both its value and its normal derivative at the screen, given by

$$\phi_\pm(\mathbf{p}) = \frac{1}{2}\left(\frac{k}{2\pi n}\right)^{N/2} \int_{\Re^N} d^N\mathbf{q} \left[\sqrt{1-p^2/n^2}\,\widetilde{\phi}(\mathbf{q}) \pm \frac{1}{ik}\widetilde{\phi}'(\mathbf{q})\right] \times \exp(-ik\mathbf{p}\cdot\mathbf{q}/n). \tag{10.4}$$

The Parseval relation between the $\mathcal{L}_2(\mathcal{S}_N)$ inner product (9.2*f*) over the full sphere defines a *non-local* inner product over the screen involving again both the wavefunction and its normal derivative thus [10]:

$$(\phi,\psi)_{\mathcal{L}_2(\mathcal{S}_N)} = \int_{\mathcal{S}_N} d^N S(\vec{p})\, \phi(\vec{p})^* \psi(\vec{p}) \tag{10.5a}$$

$$\begin{aligned} &= \left(\frac{k}{2\pi n}\right)^N \int_{\substack{\mathbf{p}\in\Re^N \\ |\mathbf{p}|<n}} \frac{d^N\mathbf{p}}{\sqrt{1-p^2/n^2}} \\ &\quad\times \int_{\Re^N} d^N\mathbf{q} \int_{\Re^N} d^N\mathbf{q}'\, \exp(-ik\mathbf{p}\cdot(\mathbf{q}-\mathbf{q}')/n) \\ &\quad\times \left[(1-p^2/n^2)\widetilde{\phi}(\mathbf{q})^*\widetilde{\psi}(\mathbf{q}') + \frac{1}{k^2}\widetilde{\phi}'(\mathbf{q})^*\widetilde{\psi}'(\mathbf{q}')\right] \\ &= \left(\frac{k}{2\pi n}\right)^2 \int_{\Re^N} d^N\mathbf{q} \int_{\Re^N} d^N\mathbf{q}' \\ &\quad\times \Big[\omega(|\mathbf{q}-\mathbf{q}'|)\widetilde{\phi}(\mathbf{q})^*\widetilde{\psi}(\mathbf{q}') \\ &\qquad + \varpi(|\mathbf{q}-\mathbf{q}'|)\widetilde{\phi}'(\mathbf{q})^*\widetilde{\psi}'(\mathbf{q}')\Big] \end{aligned} \tag{10.5b}$$

$$= (\widetilde{\phi},\widetilde{\psi})_{\mathcal{H}_k^N}. \tag{10.5c}$$

The integral over the sphere $\mathcal{S}_N$ has been performed to yield two nonlocal weight functions, ω and ϖ, that are functions of the norm of the coordinate vector difference $|\mathbf{q}-\mathbf{q}'|$

$$\begin{aligned} \omega(|\mathbf{q}-\mathbf{q}'|) &= \frac{1}{2}\int_{\mathcal{S}_N} d^N S(\vec{p})\,(1-p^2/n^2)\,\exp(-ik\mathbf{p}\cdot(\mathbf{q}-\mathbf{q}')/n) \\ &= \frac{1}{2}\int_0^n p^{N-1}\,dp\,\sqrt{1-p^2/n^2} \\ &\quad\times \int_{\mathcal{S}_{N-1}} d\omega_{\mathbf{p}}\,\exp(-ikp\,|\mathbf{q}-\mathbf{q}'|\cos[\angle(\mathbf{p},\mathbf{q}-\mathbf{q}')/n] \\ &= \sqrt{\frac{\pi}{2}}\left(\frac{k\,|\mathbf{q}-\mathbf{q}'|}{2\pi n}\right) \\ &\quad \int_0^n p^{N/2}dp\,\sqrt{1-p^2/n^2}\,J_{N/2-1}(kp\,|\mathbf{q}-\mathbf{q}'|/n) \end{aligned}$$

$$= \tfrac{1}{4}(2\pi)^{(N+1)/2} n^N \frac{J_{(N+1)/2}(k\,|\mathbf{q}-\mathbf{q}'|)}{(k\,|\mathbf{q}-\mathbf{q}'|)^{(N+1)/2}}. \qquad (10.6)$$

The integral ([21], Eq. 6.567.1) leads to a Bessel function of the first kind $J_m(x)$. Similarly, we find

$$\varpi(|\mathbf{q}-\mathbf{q}'|) = \frac{1}{2k^2}\int_{S_N} d^N S(\vec{p})\, \exp(-ik\mathbf{p}\cdot(\mathbf{q}-\mathbf{q}')/n)$$

$$= \tfrac{1}{4}(2\pi)^{(N+1)/2} n^N \frac{1}{k^2}\frac{J_{(N-1)/2}(k\,|\mathbf{q}-\mathbf{q}'|)}{(k\,|\mathbf{q}-\mathbf{q}'|)^{(N-1)/2}}. \qquad (10.7)$$

These equations may be checked with the results for $N = 1$ in Ref. [22], and for the case of ordinary optics $N = 2$ with Eqs. (10.10)–(10.13) in Ref. [10]. They involve the pleasant function $J_m(x)/x^m$ whose value at $x = 0$ is a maximum, $[2^m\Gamma(m+1)]^{-1}$; it decreases to its first zero[11] at $2.405, \pi, 3.832, 4.493$ for $m = 0, \frac{1}{2}, 1, \frac{3}{2}$ (for $N = 1$ dimension we need $m = 0$ and 1, for the case of ordinary optics $N = 2$, $m = \frac{1}{2}$ and $\frac{3}{2}$); thereafter it oscillates and decreases as $\sqrt{2/\pi}\,x^{-m-1}$ asymptotically. This is the picture of the nonlocality of the Helmholtz-Hilbert space $\mathcal{H}_k^N$ on the N-dimensional screen. The $\mathcal{H}_k^N$ space of functions does not support Dirac δ's as valid objects: images cannot be perfectly pointlike since such need $\mathbf{p} \in \Re^N$, and momentum ranges only on the subset $p < n$. The position coordinate is *not* a well defined concept in Euclidean optics as it is in Heisenberg-Weyl quantum mechanics.

The face of contact of Euclidean and Heisenberg-Weyl phase spaces is in the momentum subspace. *Plane waves* are well defined objects in both. With 'forward Heisenberg-Weyl direction' $\mathbf{P}_0 \in \Re^N$, the wave will appear in the $\mathbf{P}$-representation as

$$\Phi_{\mathbf{P}_0}(\mathbf{P}) = \delta^N(\mathbf{P}-\mathbf{P}_0), \qquad (10.8a)$$

and on the $\mathbf{Q}$-screen as the complex amplitude (10.1a),

$$\widetilde{\Phi}_{\mathbf{P}_0}(\mathbf{Q}) = \left(\frac{k}{2\pi n}\right)^{N/2} \exp(i\,\mathbf{P}_0\cdot\mathbf{Q}). \qquad (10.8b)$$

This is a periodic function with wavelength $\lambda = 2\pi n/kP_0$ between crests. The wavelength may range from infinity ($\mathbf{P}_0 = \mathbf{0}$, a ray paralell to the optical axis, its wavefronts paralell to the screen), through decreasing real values for increasing ray angles, down to zero for $P_0 \to \infty$ (this limit is nonrelativistic-mechanical, not optical).

---

[11] Except for π, the numbers are truncated to three decimals.

The (inverse) opening coma transformation asigns to $\Phi_{\mathbf{P}_0}$ through (9.4a) a Euclidean plane wave in momentum representation

$$\begin{aligned} \phi_{\mathbf{p}_0}(\mathbf{p}) &= |1-p^2/n^2|^{-(N+1)/4}\Phi_{\mathbf{P}_0(\mathbf{p}_0)}(\mathbf{P}(\mathbf{p})) \\ &= |1-p^2/n^2|^{-(N+1)/4}\delta^N\left(\frac{\mathbf{p}}{\sqrt{1-p^2/n^2}}-\frac{\mathbf{p}_0}{\sqrt{1-p_0^2/n^2}}\right) \\ &= |1-p^2/n^2|^{(N+3)/4}\delta^N(\mathbf{p}-\mathbf{p}_0). \end{aligned} \tag{10.9a}$$

This yields through (10.3b) that the Helmholtz solution on the screen is a plane wave in the Euclidean direction $\mathbf{p}_0 = \mathbf{P}_0/\sqrt{1+P_0^2/n^2}$,

$$\widetilde{\phi}_{\mathbf{p}_0}(\mathbf{q}) = \left(\frac{k}{2\pi n}\right)^{N/2}(1-p_0^2/n^2)^{(N+1)/4}\exp(ik\mathbf{p}_0\cdot\mathbf{q}/n). \tag{10.10a}$$

With the assumption that our Heisenberg-Weyl plane wave is in the *forward* hemisphere of ray directions, the normal derivative is given to us by (10.3d), and is

$$\begin{aligned} \widetilde{\phi}'_{\mathbf{p}_0}(\mathbf{q}) &= \left(\frac{k}{2\pi n}\right)^{N/2}(1-p_0^2/n^2)^{(N+3)/4}\exp(ik\mathbf{p}_0\cdot\mathbf{q}/n) \\ &= ik\sqrt{1-p_0^2/n^2}\,\widetilde{\phi}_{\mathbf{p}_0}(\mathbf{q}). \end{aligned} \tag{10.10b}$$

A backward-directed wave in Helmholtz optics has the opposite sign of the normal derivative. Linear combinations of forward and backward rays allows for *real* (or purely imaginary) standing waves throughout space, and permits the decoupling of the function from its normal derivative. This freedom can not be obtained in Heisenberg-Weyl optics for a single $\Re^N$ screen; there, we can at most linearly combine waves $\mathbf{P}_0$ and $-\mathbf{P}_0$ with real results on the screen, but complex oscillation elsewhere.

In the last formula there is a decrease factor of $(1-p_0^2/n^2)^{(N+1)/4}$ in the amplitude of the wave, which drops to zero for $p_0 = n$. A grid of Heisenberg-Weyl plane wave vectors in $\Re^N$ will map on a 'noncartesian' grid in the Descartes sphere in Escher-like distortion; as we draw near the equator the intensities will drop to zero.

Euclidean plane waves that are Dirac-normalized on $\mathcal{S}_N$ may be constructed from (10.3). These are [10]:

$$\widetilde{w}_{\mathbf{p}_0,\tau}(\mathbf{q}) = \left(\frac{k}{2\pi n}\right)^{N/2}\exp(ik\mathbf{p}_0\cdot\mathbf{q}), \tag{10.11a}$$

$$\widetilde{w}'_{\mathbf{p}_0,\tau}(\mathbf{q}) = ik\tau\left(\frac{k}{2\pi n}\right)^{N/2}\sqrt{1-p_0^2/n^2}\,\exp(ik\mathbf{p}_0\cdot\mathbf{q}). \tag{10.11b}$$

Their normalization is evident on momentum space

$$w_{\mathbf{p}_0,\tau}(\mathbf{p}) = \sqrt{1-p_0^2/n^2}\,\delta^N(\mathbf{p}-\mathbf{p}_0)\,\delta_{\tau,\mathrm{sign}\surd} = \delta^N_{\mathcal{S}_N}(\vec{p}-\vec{p}_0). \tag{10.11c}$$

In the corresponding Heisenberg-Weyl direction $\mathbf{P}_0 = \mathbf{p}_0/\sqrt{1-p_0^2/n^2}$, for the corresponding value of τ, the waves are

$$W_{\mathbf{P}_0}(\mathbf{P}) = (1 + P_0^2/n^2)^{(N+1)/4}\,\delta^N(\mathbf{P} - \mathbf{P}_0), \quad (10.12a)$$

$$\widetilde{W}_{\mathbf{P}_0}(\mathbf{Q}) = \left(\frac{k}{2\pi n}\right)^{N/2} (1 + P_0^2/n^2)^{(N+1)/4} \exp(ik\mathbf{P}_0\cdot\mathbf{Q}). \quad (10.12b)$$

Let us now consider Heisenberg-Weyl Dirac δ's on the screen, at points $\mathbf{Q}_0$, in their position and momentum representations,

$$\widetilde{D}_{\mathbf{Q}_0}(\mathbf{Q}) = \delta^N(\mathbf{Q} - \mathbf{Q}_0), \quad (10.13a)$$

$$D_{\mathbf{Q}_0}(\mathbf{P}) = \left(\frac{k}{2\pi n}\right)^{N/2} \exp(-ik\mathbf{P}\cdot\mathbf{Q}_0/n). \quad (10.13b)$$

The coma map (10.9) leads to the Euclidean wavefunction on the right momentum range:

$$d_{\mathbf{Q}_0}(\mathbf{p}) = \left(\frac{k}{2\pi n}\right)^{N/2} \frac{1}{(1 - p^2/n^2)^{(N+1)/4}} \exp -i\frac{k}{n}\left(\frac{\mathbf{p}\cdot\mathbf{Q}_0}{\sqrt{1-p^2/n^2}}\right). \quad (10.14)$$

Finally, the image on the Euclidean screen is found through the integral in (10.3) to be

$$\tilde{d}_{\mathbf{Q}_0}(\mathbf{q}) = \left(\frac{k}{2\pi n}\right)^N \int_{\Re^N} \frac{d^N\mathbf{p}}{(1 - p^2/n^2)^{(N+3)/4}} \exp i\frac{k}{n}\left(\mathbf{p}\cdot\mathbf{q} - \frac{\mathbf{p}\cdot\mathbf{Q}_0}{\sqrt{1-p^2/n^2}}\right). \quad (10.15)$$

This integral, when and if evaluated, will be the kernel of the unitary operator of symplectic coma transformation between $\mathcal{L}_2(\Re^N)$ and $\mathcal{H}_k^N$. If $\widetilde{F}(\mathbf{Q})$ is a wavefunction calculated to carry an image in Heisenberg-Weyl optics, its actual image on a Euclidean screen will be

$$\tilde{f}(\mathbf{q}) = \int_{\Re^N} d^N\mathbf{Q}\, \tilde{d}_{\mathbf{Q}}(\mathbf{q})\, \widetilde{F}(\mathbf{Q}). \quad (10.16)$$

This integral is *not quite* a convolution because the kernel $\tilde{d}_{\mathbf{Q}}(\mathbf{q})$ is *not quite* a function of $\mathbf{q} - \mathbf{Q}$.

When only rays at small angles to the optical axis are involved, then both momentum representation wavefunctions $F(\mathbf{P})$ and $f(\mathbf{p})$ are significantly different from zero only at a small neighborhood of $\mathbf{p} = \mathbf{0}$. The integral in (10.15) is then approximated with the integrand $\exp ik\,\mathbf{p}\cdot(\mathbf{q} - \mathbf{Q}_0)$, both in company with this $f(\mathbf{p})$. In the small wavelength limit ($k \to \infty$), the exponential oscillates everywhere except at $\mathbf{q} = \mathbf{Q}_0$ where it diverges as a Dirac δ in $\Re^N$. For small ray angles thus, Euclidean images are sharp, with no magnification or distortion. For realistic ranges of $\mathbf{P}$, the integral in (10.15) should be interesting to study as it approximates also the *Airy* function when the next higher term in the exponent is computed.

7.11 Gaussians, non-diffracting beams, and concluding remarks

Coherent states, correlated (*squeezed*) and discrete [4], and generally Gaussian beams [1], are prime fruit of the Heisenberg-Weyl model of optics. This is so because we may *complexify* the inhomogeneous linear symplectic transformations. Imaginary-z free propagation, generated by P^2, produces Gaussians out of the Dirac δ in $\Re^N$, just as heat diffusion does. The $\Re^N$-Fourier transform of a Gaussian is generically a Gaussian. As operators, neither $\mathbf{Q}$ nor $\mathbf{P}$ have sharp eigenvalues in this set of functions, but their dispersions satisfy the minimum of the Heisenberg uncertainty relation.

Through the coma map (9.4), the Heisenberg-Weyl Gaussians with exponents $-AP^2$ ($A > 0$) and/or $\mathbf{B}\cdot\mathbf{P}$, will become exponentials of $-Ap^2/(1-p^2/n^2)$ and $\mathbf{B}\cdot\mathbf{p}/\sqrt{1-p^2/n^2}$, times the measure factor $(1-p^2/n^2)^{(N+1)/4}$. These functions, although not *quite* Gaussians (because their exponent is not simply $-ap^2$ and $\mathbf{B}\cdot\mathbf{p}$), do adequately go to zero at $|\mathbf{p}| \to n$; *i.e.*, approaching the *surface* of the solid N-sphere of Euclidean momenta. In the ordinary case $N = 2$, this is the 1-dimensional *rim* of the forward momentum disk, the equatorial circle. Since only forward (or backward) rays are involved, the Helmholtz solution and its normal derivative at the screen are related by a factor of $\pm ik\sqrt{1-p^2/n^2}$, as was the case with plane waves in (10.10*b*). The $\mathcal{H}_k^N$ and $\mathcal{L}_2(\mathcal{S}_N)$ inner product properties of these projected Heisenberg-Weyl coherent states will be the same as their well-known counterparts in $\mathcal{L}_2(\Re^N)$ because of the Parseval relation (9.2).

What is the natural counterpart of a Gaussian distribution on the surface of the Descartes sphere of ray directions $\mathcal{S}_N$? Take a Dirac δ in the context of $\mathcal{L}_2(\mathcal{S}_N)$ and let it *diffuse*, as if it were heat, with $\exp -w\Delta_{\mathcal{S}_N}$, where $\Delta_{\mathcal{S}_N}$ is the Laplace-Beltrami operator on $\mathcal{S}_N$. For small width w the Descartes $\mathcal{S}_N$-Gaussian looks like a Heisenberg-Weyl $\Re^N$-Gaussian centered on the original δ, but as w increases it asymptotically tends to a nonzero constant over the finite volume of $\mathcal{S}_N$ and both momentum charts are necessarily involved. For the case of $N = 1$ dimensional screens, we can point to Ref. [17] to show that this Gaussian is the heat kernel on a conducting ring $\mathcal{S}_1$ —one of the Jacobi ϑ-functions. The Fourier *series* decomposition of $\mathcal{L}_2(\mathcal{S}_1)$ yields naturally a description of the images on the screen in terms of equally-spaced *sampling points*; that is unitarily equivalent to the Hilbert space $\mathcal{H}_k^1$. The Euclidean Gaussian in this discrete image space is a *modified* Bessel function $I_j(kw)$ that, on the *integers* j, is symmetric about a smooth peak and falls off asymptotically ($|j| \to \infty$) as an ordinary Gaussian of width w does. This takes us to the theory of coherent states on group manifolds and their sampling theorem, that we hope to develop elsewhere.

Among the subgroup-adapted wavefunction bases we should draw also special attention to the *diffraction-free* beams [23]. We may define these to be the Euclidean wavefunction, p–$\omega_{\mathbf{p}}$ separated solutions of the Helmholtz equation on the screen, that are eigenfunctions of the generator of translations along the $(N+1)^{\text{th}}$ axis (the optical z-axis), namely h^{τ} in (7.1*b*). We can treat the two charts separately and work only with *forward* beams; this fixes the normal derivative of the solution as for plane waves in (10.10*b*). The solution of $h\phi_a = \lambda(a)\phi_a$ in $\mathbf{p}$-space is a Dirac δ with support on a single value of the radius $|\mathbf{p}| = p = a$, times a function $\Upsilon(\omega_{\mathbf{p}})$ of the $\mathcal{S}_{N-1}$ angular variables in $\mathbf{p} = p\omega_{\mathbf{p}}$, times a normalization constant $\nu_E(a)$. In the ordinary $N = 2$ case the Dirac δ has support on a circle inside the momentum disk. In Refs. [23] a mask with an annular slit provided the source for a ray distribution that, after a Fourier transformer lens, provided a *diffraction-free* or J_0 beam, with a Bessel function amplitude on successive z-translated screens.[12]

In the N-dimensional case, a momentum wavefunction $\phi_a(p, \omega_{\mathbf{p}}, \tau = +)$ that contains $\delta(p - a)$ will cast an image $\widetilde{\phi}_a(\mathbf{q})$ given by (10.3*a*, *b*). The steps needed to decompose the integration into radius p and $\mathcal{S}_{N-1}$-angles $\omega_{\mathbf{p}}$ follow closely the integration in (10.6), except that the integrand factor $(1 - p^2/n^2)$, is replaced by $\delta(p - a)$, and we have plain $\mathbf{q}$ here instead of $(\mathbf{q} - \mathbf{q}')$. The angular integral over $\mathcal{S}_{N-1}$ now contains the angular function $\Upsilon(\omega_{\mathbf{p}})$. If this is taken to be a constant, the diffraction-free beam $\widetilde{\phi}_a(\mathbf{q})$ will have the form of a Bessel function $J_{N/2-1}(ka|\mathbf{q}|/n)$. In $N = 2$ dimensions this is the J_0 beam of reference [23].

If the angular function $\Upsilon(\omega_{\mathbf{p}})$ is *not* a constant, the diffraction-free beam $\widetilde{\phi}_a(\mathbf{q})$ will also exhibit angular dependence. In the ordinary case $N = 2$ this leads to nondiffracting solutions $e^{im\omega} J_{|m|}(kaq/n)$, where $m = 0, \pm 1, \pm 2, \ldots$ is the in-screen rotations' SO(2) irreducible representation label, and ω the single polar angle. For general N we may use SO(N) spherical harmonics as a basis where the vectors are classified by a complete set of labels following $\mathrm{SO}(N) \supset \mathrm{SO}(N-1) \supset \cdots \supset \mathrm{SO}(2)$. This set of labels will replace the single m. The SO(N) Casimir is found as $\frac{1}{2}\sum_{i,j}(R_{i,j})^2$ from (7.2*a*), corresponding to the well-known Petzval invariant $(\mathbf{p} \times \mathbf{q})^2$ in the $N = 2$ case. While $\mathbf{p} \times \mathbf{q}$ by itself yields m, it is the Casimir eigenvalue $\kappa^{\mathrm{SO}(N)} = \ell(\ell + N - 2)$, $\ell = 0, 1, 2, \ldots$, that will appear in the radial part as ℓ in $J_{N/2+\ell-1}(ka|\mathbf{q}|/n)$ for the general-N case.

---

[12]The very interesting experimental property of such beams is that, due to the finiteness of the wavefront, the beam mantains its peaked J_0 shape up to a certain distance only; thereafter, the peak vanishes rather abruptly. Very much as Luke Skywalker's 3-foot-long red laser sword, and Darth Vader's blue one. (This would *still* not explain why such light swords would clash among themselves, unless a nonlinear régime is encountered.)

The Euclidean $q_{N+1} = z$-evolution is produced by $\exp(ikz\hat{h}/n)$. On its eigenfunction $\widetilde{\phi}_a(\mathbf{q})$ this yields a phase of $\exp(ikz\sqrt{1-a^2/n^2})$ where a is, as before, the direction $p = n \sin\theta_a$, $\theta_a = \arcsin(a/n) \leq \frac{1}{2}\pi$, of the constitutent forward plane waves that form the nondiffracting beam. Finally, the normal derivative $\widetilde{\phi}'_a(\mathbf{q}, z)$ is $ik\sqrt{1-a^2/n^2}$ times $\widetilde{\phi}_a(\mathbf{q}, z)$.

Bessel functions of the first kind *also* appear in Heisenberg-Weyl optics, as was pointed out in Ref. [24] for the $N = 2$ case of 'physical' optics. Indeed, the subgroup reduction $\mathrm{Sp}(2N, \Re) \supset \mathrm{Sp}(2, \Re) \otimes \mathrm{SO}(N)$ in the oscillator realization has the same $\mathrm{SO}(N)$ generators $R_{i,j}$ of the Euclidean group (7.2a), the $\mathrm{Sp}(2,\Re)$ generators in (6.3), and the subgroups are *conjugate*, *i.e.*, their Casimir operators are related by $\kappa^{\mathrm{Sp}(2)} = -\frac{1}{4}\kappa^{\mathrm{SO}(N)} + \frac{1}{16}N(4-N)$. The value of the Bargmann label of $\mathrm{Sp}(2,\Re)$ is $\frac{1}{2}(\ell + \frac{1}{2}N)$. We look at the *eigenbasis* of $\frac{1}{2}P^2$, the generator of a *parabolic* orbit in $\mathrm{Sp}(2,\Re)$ [25] in the Q-realization. The eigenfunction corresponding to eigenvalue $\frac{1}{2}A^2 \in \Re^+$ contains a Bessel factor of $J_{N/2+\ell-1}(kA|\mathbf{Q}|/n)$, an $\mathcal{L}_2(\Re^N)$ normalization factor $\nu_R(A)$, and an $\mathrm{SO}(N) \subset \mathrm{SO}(N-1) \subset \ldots \subset \mathrm{SO}(2)$ set of labels for the angular part, just like their Euclidean counterparts, but for $A = a/\sqrt{1-a^2/n^2}$ now unbounded. The forward z-evolution is in this context the *paraxial* one, *i.e.*, generated by $H = P^2/2n$. This will produce on the wavefunction a phase of $\exp(-ikzA^2/2n^2)$. Comparing this with the phase $\exp(ikz\sqrt{1-a^2/n^2}) = \exp ikz\ \exp(-ikza^2/2n^2) \times \exp(-ikza^4/8n^4)\cdots$ obtained for Euclidean propagation, we see that the Heisenberg-Weyl wavefunction still needs the well-known (central) phase $\exp ikz$, and approximates well the phase anomaly $-ika^2/2n^2$ for $A \approx a \ll n$. It was the point of Ref. [25] to show that the Heisenberg-Weyl nondiffracting beams, subject to arbitrary paraxial transformations, generically mantain their dependence on a Bessel function (scaled by z and/or multiplied by an oscillating Gaussian), *i.e.*, the beams *self-reproduce* under the group of paraxial transformations.

The above two classes of beams have been given as illustrations of the opening coma map between Euclidean and Heisenberg-Weyl wave optics. The latter has served for many developments that still await to be defined on the former. Signal theory, in particular its formulation through the Wigner phase space distribution function and the Gabor expansion [26], do not seem applicable as such to wide-angle wave optics. For example, we may inquire about the optimal sampling-point set for Helmholtz solutions for 4π or stopped beams, with or without axial symmetry, or about the natural spread of coherent states and the behaviour of their correlation (*squeezing*). The purpose of illustration is served by leaving these subjects for further development elsewhere.

7.12 Acknowledgements

We are in debt with François Leyvraz and Thomas H. Seligman, Instituto de Física, Cuernavaca, for many interesting discussions on the comatic phenomenon in phase space. One of us (V.I.M.) wishes to thank the **Instituto de Investigaciones en Matemáticas Aplicadas y en Sistemas, UNAM**, for the excellent working conditions in the Cuernavaca Unit.

Dr. Man'ko's visit to Mexico was made possible by the bilateral academic exchange agreement between the **USSR Academy of Sciences** and the **Consejo Nacional de Ciencia y Tecnología**. He was DISTINGUISHED VISITOR of the CENTRO INTERNACIONAL DE FÍSICA Y MATEMÁTICAS APLICADAS AC. The donor of the Visitors program fund was the **Fondo de Fomento Educativo BCH**, whose generous support of scientific endeavour we acknowledge with pleasure.

7.13 REFERENCES

[1] V.I. Man'ko and K.B. Wolf, The influence of spherical aberration on gaussian beam propagation. In *Lie Methods in Optics*, ed. by J. Sánchez-Mondragón and K.B. Wolf. Lecture Notes in Physics, Vol. 250 (Springer Verlag, Heidelberg, 1986); K.B. Wolf and V.I. Man'ko, *ibid.* [in Russian]. Trudy Fiz. Inst. P.N. Lebedev, Vol. **176** (Nauka, Moscow, 1986); translated in *Classical and Quantum Effects in Electrodynamics*, ed. by A.A. Komar (Nova Science Publ., Commack N.Y., 1988), pp. 169–200.

[2] R. Glauber, Coherent states of the quantum oscillator, *Phys. Rev. Lett.* **10**, 84 (1963).

[3] *See for example*, M.M. Nieto, What are squeezed states really like? In *Frontiers of Nonequilibrium Statistical Physics*, ed. by G.T. Moore and M.O. Scully (Plenum Publ. Corp, New York, 1986), pp. 287–307.

[4] V.V. Dodonov, E.V. Kurmyshev, and V.I. Man'ko Exact bounds for the uncertainty relation in correlated coherent states [in Russian]. In *Group-theoretical Methods in Physics*, Proceedings of the Zvenigorod Seminar, November 1979. (Nauka, Moscow, 1980).

[5] V.V. Dodonov, E.V. Kurmyshev, and V.I. Man'ko Generalized uncertainty relation and correlated coherent states. *Phys. Lett.* **A79**, 150–152 (1980).

[6] A.J. Dragt, E. Forest, and K.B. Wolf, Foundations of a Lie algebraic theory of geometrical optics. In *Lie Methods in Optics*, *op. cit.*

[7] K.B. Wolf, The Heisenberg–Weyl ring in quantum mechanics. In *Group Theory and its Applications*, Vol. III, ed. by E.M. Loebl (Academic Press, New York, 1975), pp. 189–247.

[8] H. Raszillier and W. Schempp, Fourier optics from the perspective of the Heisenberg group. In *Lie Methods in Optics, op. cit.*

[9] J.A. Arnaud, *Beam and Fiber Optics* (Academic Press, New York, 1976).

[10] K.B. Wolf, Elements of Euclidean Optics. In this volume.

[11] K.B. Wolf, Symmetry in Lie optics, *Ann. Phys.* **172**, 1–25 (1986).

[12] N.M. Atakishiyev, W. Lassner, and K.B. Wolf, The relativistic coma aberration. I. Geometrical optics. *Comunicaciones Técnicas IIMAS* No. 509 (1988); *ib.* II. Helmholtz wave optics. No. 517 (1988), to appear in *Journal of Mathematical Physics.*

[13] T. Sekiguchi and K.B. Wolf, The Hamiltonian formulation of optics, *Am. J. Phys.* **55**, 830–835 (1987).

[14] K.B. Wolf, Nonlinearity in aberration optics. In *Symmetries and Nonlinear Phenomena*, Proceedings of the International School on Applied Mathematics, Centro Internacional de Física, Paipa, Colombia, 22–26 Feb. 1988, ed. by D. Levi and P. Winternitz, CIF Series Vol. 9 (World Scientific, Singapore, 1989).

[15] O.N. Stavroudis, *The Optics of Rays, Wavefronts, and Caustics* (Academic Press, New York, 1972); Eq. (II-19) on p. 26.

[16] S. Steinberg, Lie series, Lie transformations, and their applications. In *Lie Methods in Optics, op. cit.*; pp. 45–103.

[17] K.B. Wolf, A Euclidean algebra of Hamiltonian observables in Lie optics, *Kinam* **6**, 141–156 (1985).

[18] A.J. Dragt, A Lie algebraic theory of geometrical optics and optical aberrations. *J. Opt. Soc. Am.* **72**, 372–379 (1982).

[19] J.R. Klauder, Wave theory of imaging systems. In *Lie methods in optics, op. cit.* pp. 183–191; Eqs. (11) on p. 186.

[20] F. Leyvraz and T.H. Seligman, Sequences of point transformations and linear canonical transformations in classical and quantum mechanics. Preprint IFUNAM (1988), to appear in *J. Phys.*

[21] I.S. Gradshteyn and I.M. Ryzhik, *Tables of integrals, sums, series and products* (Academic Press, New York, 1975).

[22] S. Steinberg and K.B. Wolf, Invariant inner products on spaces of solutions of the Klein-Gordon and Helmholtz equations, *J. Math. Phys.* **22**, 1660–1663 (1981).

[23] J. Durnin, J.J. Miceli, and J.H. Eberly, Diffraction-free beams. *Phys. Rev. Lett.* **58**, 1499–1501 (1987); J. Durnin, Exact Solutions for non-diffracting beams. I. The scalar theory. *J. Opt. Soc. Am. A* **4**, 651–654 (1987).

[24] K.B. Wolf, Diffraction-free beams remain diffraction-free under all paraxial optical transformations. *Phys. Rev. Lett.* **60**, 757–759 (1988).

[25] D. Basu and K.B. Wolf, The unitary irreducible representations of *SL(2,R)* in all subgroup reductions. *J. Math. Phys.* **23**, 189–205 (1982).

[26] M.J. Bastiaans, *Local-Frequency Description of Optical Signals and Systems*, Eindhoven University of Technology report 88-E-191 (April, 1988); lectures delivered at the First International School and Workshop on Photonics (Oaxtepec, June 28 – July 8, 1988).

Lecture Notes in Mathematics

Vol. 1236: Stochastic Partial Differential Equations and Applications. Proceedings, 1985. Edited by G. Da Prato and L. Tubaro. V, 257 pages. 1987.

Vol. 1237: Rational Approximation and its Applications in Mathematics and Physics. Proceedings, 1985. Edited by J. Gilewicz, M. Pindor and W. Siemaszko. XII, 350 pages. 1987.

Vol. 1250: Stochastic Processes – Mathematics and Physics II. Proceedings 1985. Edited by S. Albeverio, Ph. Blanchard and L. Streit. VI, 359 pages. 1987.

Vol. 1251: Differential Geometric Methods in Mathematical Physics. Proceedings, 1985. Edited by P.L. García and A. Pérez-Rendón. VII, 300 pages. 1987.

Vol. 1255: Differential Geometry and Differential Equations. Proceedings, 1985. Edited by C. Gu, M. Berger and R.L. Bryant. XII, 243 pages. 1987.

Vol. 1256: Pseudo-Differential Operators. Proceedings, 1986. Edited by H.O. Cordes, B. Gramsch and H. Widom. X, 479 pages. 1987.

Vol. 1258: J. Weidmann, Spectral Theory of Ordinary Differential Operators. VI, 303 pages. 1987.

Vol. 1260: N.H. Pavel, Nonlinear Evolution Operators and Semigroups. VI, 285 pages. 1987.

Vol. 1263: V.L. Hansen (Ed.), Differential Geometry. Proceedings, 1985. XI, 288 pages. 1987.

Vol. 1265: W. Van Assche, Asymptotics for Orthogonal Polynomials. VI, 201 pages. 1987.

Vol. 1267: J. Lindenstrauss, V.D. Milman (Eds.), Geometrical Aspects of Functional Analysis. Seminar. VII, 212 pages. 1987.

Vol. 1269: M. Shiota, Nash Manifolds. VI, 223 pages. 1987.

Vol. 1270: C. Carasso, P.-A. Raviart, D. Serre (Eds.), Nonlinear Hyperbolic Problems. Proceedings, 1986. XV, 341 pages. 1987.

Vol. 1272: M.S. Livšic, L.L. Waksman, Commuting Nonselfadjoint Operators in Hilbert Space. III, 115 pages. 1987.

Vol. 1273: G.-M. Greuel, G. Trautmann (Eds.), Singularities, Representation of Algebras, and Vector Bundles. Proceedings, 1985. XIV, 383 pages. 1987.

Vol. 1275: C.A. Berenstein (Ed.), Complex Analysis I. Proceedings, 1985–86. XV, 331 pages. 1987.

Vol. 1276: C.A. Berenstein (Ed.), Complex Analysis II. Proceedings, 1985–86. IX, 320 pages. 1987.

Vol. 1277: C.A. Berenstein (Ed.), Complex Analysis III. Proceedings, 1985–86. X, 350 pages. 1987.

Vol. 1283: S. Mardešić, J. Segal (Eds.), Geometric Topology and Shape Theory. Proceedings, 1986. V, 261 pages. 1987.

Vol. 1285: I.W. Knowles, Y. Saitō (Eds.), Differential Equations and Mathematical Physics. Proceedings, 1986. XVI, 499 pages. 1987.

Vol. 1287: E.B. Saff (Ed.), Approximation Theory, Tampa. Proceedings, 1985–1986. V, 228 pages. 1987.

Vol. 1288: Yu. L. Rodin, Generalized Analytic Functions on Riemann Surfaces. V, 128 pages, 1987.

Vol. 1294: M. Queffélec, Substitution Dynamical Systems – Spectral Analysis. XIII, 240 pages. 1987.

Vol. 1299: S. Watanabe, Yu.V. Prokhorov (Eds.), Probability Theory and Mathematical Statistics. Proceedings, 1986. VIII, 589 pages. 1988.

Vol. 1300: G.B. Seligman, Constructions of Lie Algebras and their Modules. VI, 190 pages. 1988.

Vol. 1302: M. Cwikel, J. Peetre, Y. Sagher, H. Wallin (Eds.), Function Spaces and Applications. Proceedings, 1986. VI, 445 pages. 1988.

Vol. 1303: L. Accardi, W. von Waldenfels (Eds.), Quantum Probability and Applications III. Proceedings, 1987. VI, 373 pages. 1988.

Lecture Notes in Physics

Vol. 329: A. Heck, F. Murtagh (Eds.), Knowledge Based Systems in Astronomy. IV, 280 pages. 1989.

Vol. 330: J.M. Moran, J.N. Hewitt, K.Y. Lo (Eds.), Gravitational Lenses. Proceedings, 1988. XIV, 238 pages. 1989.

Vol. 331: G. Winnewisser, J.T. Armstrong (Eds.), The Physics and Chemistry of Interstellar Molecular Clouds mm and Sub-mm Observations in Astrophysics. Proceedings, 1988. XVIII, 463 pages. 1989.

Vol. 332: P. Flin, H.W. Duerbeck (Eds.), Morphological Cosmology. Proceedings, 1988. VII, 438 pages. 1989.

Vol. 333: I. Appenzeller, H.J. Habing, P. Léna (Eds.), Evolution of Galaxies – Astronomical Observations. Proceedings, 1988. X, 391 pages. 1989.

Vol. 334: L. Maraschi, T. Maccacaro, M.-H. Ulrich (Eds.), BL Lac Objects. XIII, 497 pages. 1989.

Vol. 335: A. Lakhtakia, V.K. Varadan, V.V. Varadan, Time-Harmonic Electromagnetic Fields in Chiral Media. VII, 121 pages. 1989.

Vol. 336: M. Müller, Consistent Classical Supergravity Theories. VI, 125 pages. 1989.

Vol. 337: A.P. Maclin, T.L. Gill, W.W. Zachary (Eds.), Magnetic Phenomena. Proceedings, 1988. VI, 142 pages. 1989.

Vol. 338: V. Privman, N.M. Švrakić, Directed Models of Polymers, Interfaces, and Clusters: Scaling and Finite-Size Properties. VI, 120 pages. 1989.

Vol. 339: F. Ehlotzky (Ed.), Fundamentals of Laser Interactions II. Proceedings, 1989. XI, 317 pages. 1989.

Vol. 340: M. Peshkin, A. Tonomura, The Aharonov-Bohm Effect. VI, 152 pages. 1989.

Vol. 341: E.F. Milone (Ed.), Infrared Extinction and Standardization. Proceedings, 1988. III, 79 pages. 1989.

Vol. 342: M. Balabane, P. Lochak, C. Sulem (Eds.), Integrable Systems and Applications. Proceedings, 1988. VII, 342 pages. 1989.

Vol. 343: M. Month, S. Turner (Eds.), Frontiers of Particle Beams; Observation, Diagnosis and Correction. Proceedings, 1988. IX, 509 pages. 1989.

Vol. 344: M. Rascle, D. Serre, M. Slemrod (Eds.), PDEs and Continuum Models of Phase Transitions. Proceedings, 1988. VII, 229 pages. 1989.

Vol. 345: H. Holden, A. Jensen (Eds.), Schrödinger Operators. Proceedings, 1988. V, 458 pages. 1989.

Vol. 346: D. Lüst, S. Theisen, Lectures on String Theory. VII, 346 pages. 1989.

Vol. 347: Ph. Blanchard, J.-P. Dias, J. Stubbe (Eds.), New Methods and Results in Non-linear Field Equations. Proceedings, 1987. VII, 133 pages. 1989.

Vol. 348: A. Bohm, M. Gadella, Dirac Kets, Gamow Vectors and Gel'fand Triplets. VII, 119 pages. 1989.

Vol. 349: Y.A. Kubyshin, J.M. Mourão, G. Rudolph, I.P. Volobujev, Dimensional Reduction of Gauge Theories, Spontaneous Compactification and Model Building. X, 80 pages. 1989.

Vol. 350: G. Tenorio-Tagle, M. Moles, J. Melnick (Eds.), Structure and Dynamics of the Interstellar Medium. Proceedings, 1989. XXI, 537 pages. 1989.

Vol. 351: A. Dervieux, B. Larrouturou (Eds.), Numerical Combustion. Proceedings, 1989. VII, 481 pages. 1989.

Vol. 352: K.B. Wolf (Ed.), Lie Methods in Optics II. Proceedings, 1988. XII, 197 pages. 1989.